Transnational Ecocinema

Transnational Ecocinema
Film Culture in an Era of Ecological Transformation

Edited by
Pietari Kääpä and Tommy Gustafsson

intellect Bristol, UK / Chicago, USA

First published in the UK in 2013 by
Intellect, The Mill, Parnall Road, Fishponds, Bristol, BS16 3JG, UK

First published in the USA in 2013 by
Intellect, The University of Chicago Press, 1427 E. 60th Street,
Chicago, IL 60637, USA

A catalogue record for this book is available from the
British Library.

Cover designer: Ellen Thomas
Copy-editor: MPS Technologies
Production manager: Jelena Stanovnik
Typesetting: Planman Technologies

ISBN 978-1-84150-729-3

Printed and bound by Hobbs, UK

Table of Contents

Acknowledgements

This volume would not have been possible without the institutions and individuals that provided both structural support and critical feedback during its construction. The collective work that has resulted in *Transnational Ecocinema* has benefitted from inspired contributions by both established and emergent scholars. This combination of perspectives appropriately reflects the emergent interdisciplinary nature of ecocinema studies. As part of a concrete effort to expand the field beyond national industries or culturally located forms of production, the global scope envisioned by the writers is reflected in their personal circumstances—we are fortunate to be able to rethink ecocinema through contributions from scholars based in different continents, from Hong Kong to the United States, from Australia to Latin America. Not only do they write in different geographical and—cultural locations, but their work engages with the ways ecological considerations extend and flow across borders. We would like to extend our gratitude to the following individuals for their comments and support on preparing the manuscript: Lars-Gustaf Andersson, Henry Bacon, Pat Brereton, Chia-Ju Chang, Sean Cubitt, Jiayan Mi, Salma Monani, Sheldon Lu, Stephen Rust, Leonie Rutherford, Deb Verhoeven, Ingrid Stigsdotter, Jiwei Xiao. Our editors at Intellect, especially Jelena Stanovnik, have provided their usual level of professionalism and support and made the process of editing the book so much more efficient. Last, we would like to thank Holger och Thyra Lauritzens stiftelse för främjande av filmhistorisk verksamhet to whom we are much obliged for their generous economic support for the publication of this book.

Acknowledgements

PART I

Introduction to Transnational Ecocinema

Introduction: Transnational Ecocinema in an age of Ecological Transformation

Pietari Kääpä and Tommy Gustafsson

Recent years have seen the exponential increase of critical books and articles on ecocinema. These range from seminal works on Hollywood cinema (Ingram 2000; Brereton 2005) to books on specific film cultures (as in the case of Lu and Mi 2009, China). Other collections take a more widespread approach (Willoquet-Maricondi 2010, issues of *International Studies in Literature and Environment*) and contribute to the ongoing proliferation of ecocinema studies. For defining the parameters of ecocinema, Lu and Mi's collection provides a concise but suggestive delineation: firstly, it is a critical grid, an interpretative strategy. Secondly, it is a description of a conscious film practice among a range of different artists and producers. Thinking of the parameters of the field in these terms opens the study to consider films from a perspective that emphasizes their ecological dimensions. It also works to encompass films that have been produced with a participatory ecological dimension in mind. As a burgeoning interdisciplinary form of film studies, ecocinema works to bring back a sense of political participation to a field that has lost some of its explicit engagement with political issues.

While film producers may aim to impart a sense of ecological responsibility to their products, what potential does cinema have for actively challenging environmental deprivation? First of all, film production is a part of the creative industries, and as a form of cultural activity that consumes considerable resources, it leaves a substantial material footprint in its wake. Producers such as Roland Emmerich and James Cameron have recently emphasized the adoption of 'green' approaches for their productions; but does this appoach sufficiently compensate for the extent of the resources demanded by the consumption, the production and distribution of films? Feature films are also often seen as forms of entertainment and, only on occasion, treated as something that takes part in social and political debates. If they do so, they may not be taken seriously as documentaries or even works of literature (fictional or not) would be. This is especially the case with Hollywood, which are often derided for their consumerist ideologies. While audiences are certainly capable of reading films like *Avatar* (2009) and *The Day After Tomorrow* (2004) as ecological, they are often only dismissed as 'movies' and not as films with an actual ecological or environmental contribution to make. Perhaps the real and most pertinent question we should ask is not how cinema can make a contribution to global ecopolitics but whether, ultimately, it can do something beyond raise awareness.

One way of attempting to confront this question directly is turning attention to transnational concerns and the approaches they entail. While prolific caricatures see the field as concerned with 'imported' films and esoteric art-house themes, this is more of a hangover of the proliferation and acceptance of Hollywood-type mainstream film culture as the global norm, which, of course, is a typical Western bound notion that neglects the huge cinematic output and distribution of films from Hong Kong, Bollywood and, in the last 20 years, Nollywood, for example. Rather than augmenting the marginalization of transnational film culture as a cultural economic other, we do not take 'transnational cinema' as 'world cinema', implying that it would be considered as Hollywood's other, or as a form of art cinema distinct from the commercial mainstream. Studies in transnational cinema are not only to do with 'art' films, but are rather a result of an increased realization of the importance of cultural flow and circulation throughout film history. They concern not only investigation of thematic influence and distribution arrangements for specific 'national' films but also increased understanding of the ways that global film markets are intertwined and coefficient.

Focusing on what the term 'transnational' implies—border crossings on a wide variety of levels—we explore its viability for ecocinema studies. Firstly, synergizing these considerations enables us to ask how a transnational scope and sense of connectivity may expand producers and audiences' ecological perception and cognitive abilities. In 'Toward an Eco-cinema', Scott MacDonald discusses experimental cinema and the challenges it provides viewers, suggesting 'the fundamental job of eco-cinema [as] retraining of perception, as a way of offering an alternative to conventional media-spectatorship' (2004: 109). This implies the need for a Brechtian challenge to spectators who are confronted with complex cinematic material that forces them to think differently and asks them to use this cognitive invigoration for politicized purposes.

Our collection does not encourage favouritism of experimental cinema as we consider a range of different types of ecocinema as capable of igniting the necessary ecocritical rethinking. For example, Roberto Forns-Broggi's chapter on political ecodocumentaries and experimental ecovideos about the Latin American conception of 'Good Life' reveals that ecocinema does not have to be 'difficult' in order to be political, and furthermore, that these ecofilms not only have the ability to provoke political thought, but also to move audiences to overcome the way in which society ignores nature.

Indeed, while these experimental films nevertheless propose a distinct formal challenge to viewers accustomed to mainstream cinema, our introduction of the 'transnational' in the transnational cinema does not have to mean difficult films. In fact, it is not the formal qualities of these films that are their most significant contribution. Instead, it is their ability to navigate between a range of different cinematic paradigms that allows them to generate complexity. They call for critical interventions that allow for multiple and contradictory meanings, even in the construction of ecological environmental rhetoric. This is a key intervention in pushing studies of ecocinema out of the somewhat simplistic endorsement of any cinematic product with a 'green' message. If ecocritical film studies want to be taken seriously they need to be prepared to be entirely critical of their own parameters.

Transnational Ecocinemas

A holistic eco-cinecriticism would closely analyze not only the representations found in a film but the telling of the film itself—its discursive and narrative structures, its inter-textual relations with the larger world, its capacities for extending or transforming perception of the larger world—and the actual contexts and effects of the film and its technical and cultural apparatus in the larger world.

(Ivakhiv 2008: 18)

Adrian Ivakhiv's call for a more penetrating and expansive form of 'eco-cinecriticism' is entirely necessary to strengthen the role of ecocinema studies in the academia. These are the types of concerns that transnational films necessarily bring to the analytical table. The process of cultural circulation—how and where the cultural products are made, what sort of content is contained in them, how they are consumed and what sort of social relations this engenders, and how they are reproduced into new material and meanings—is what interests us in this volume. To capture this complexity, different continental-geographical filmic objects stretch from Taiwan and China, Australia, Latin America, Africa and Antarctica to Europe and back to Hollywood in an effort to reveal how transnationalism can show us some of the processes through which circulation becomes ecological.

Lu and Mi's seminal collection *Chinese Ecocinema* carries a suggestive subtitle—'In an Age of Environmental Challenge'—designed to provoke readers to respond to political issues that need urgent consideration. The similarities in structure to ours—in an era of ecological transformation—are entirely intentional as we consider the present volume an expansion of work initiated by certain approaches in *Chinese Ecocinemas*. First of all, we encourage analytical frameworks beyond nations, taking into account the advantages transnational approaches bring. Secondly, our book certainly discusses environmental challenges that are increasingly becoming prevalent on a global scale. But we also emphasize that not only these challenges, but academics' responsibilities in responding to them are undergoing fundamental transformations. Thus, our collection includes both critical perspectives on ecocinema as well as explorations of what ecocritics can aim to achieve with their work. Part of this task is to respond to some of the limitations that persist in this emergent field.

Currently, much of ecocritical work on cinema is too reliant on ideological political readings of texts, which is a mode of analysis originating clearly from literature-based ecocriticism. This is understandable as ecocinema remains an emerging field. As a part of the creative industries, cinema needs to be considered in a much wider context beyond analytical readings. After all, what type of contribution can a subjective reading of a text ultimately have? While such contributions certainly add to the advancement of knowledge on ecocinema, an 'educated' reading still remains one comment amongst many generated by a given film. The contributors of this collection tackle this question head-on while providing their own voices to the discussion. To initiate this sort of rethinking, the chapters in this collection interrogate the political potential of cinema from a range of angles, including reception, distribution, production and thematic content. For example, Ines Crespo and Angela Pereira study the ways European spectators engage with diverse ecocinematic content and adopt it for a range of different purposes. Tommy Gustafsson takes an alternative angle on reception studies, choosing to focus on the ways critical prestige and ecological rhetoric operates in creating awareness around Oscar campaigns and cycles of mainstream ecocinema. Other chapters move to create new avenues for cinematic ecocriticism as they focus on transmedia and interdisciplinary concerns, as is the case with Rebecca Coyle and Susan Ward's study of transmediality in the Australian context.

Audiences, transmediality, distribution practices, documentary, fiction films, art house, and the mainstream are all part of contemporary ecocinema and exploring these areas from a transnational angle will expand the types of approaches and texts that are seen as appropriate for ecocinematic studies. The reasons for such a *modus operandi* are simple: to interrogate the participatory potential of cinema in ecological debates. In order to unravel some of the ways in which the intersections of transnationalism and ecocriticism advance both fields, our introduction will focus on case studies of recent well-known and not-so-well-known ecofilms, separated into sections on documentaries and fiction films.

Transnational Ecodocumentaries

The concept of transnational cinema, when applied to the production of documentary films, shows certain important variations from the associations created by mainstream fiction film. The documentary has not traditionally been associated with Hollywood as a concept when it comes to, for example, classical narration, subjects, or the claim that the outcome is dreamlike fabrications of 'real life' as has been the case with the feature film. Documentaries and especially ecocritical documentaries such as *Travel to Dongsha Islands* (2005), *An Inconvenient Truth* (2006), *When Clouds Clear* (2006), *Encounters at the End of the World* (2007) and *Home* (2009), all discussed in this collection, are on the contrary taken seriously and are often hailed as important regardless of their national origin or subject matter. The 'transnationality' of the ecodocumentary therefore seems to work on the level of transparency, that is, the national origin of the sender does not seem to matter if the subject is 'nature' in a wide sense. This is something that becomes even more evident when considering the main distribution channels for ecodocumentaries, namely television and the Internet.

Since the 1950s, most documentaries have been made directly for television, and those that are produced for and shown on the big screen usually get their biggest audience when they are subsequently aired on TV. BBC's heavily awarded nature/wildlife documentary series *Planet Earth* (2006) and *Life* (2009) are cases in point; the breathtaking images and the factual presentation are not aimed at a particular national body but to the world at large, as in 'it concerns us all'. This 'world level' acts as if it is neutral but in reality it is aimed at specific Western audiences, comfortably leaning back at home in their living rooms in front of their television sets. What is more, this also includes the images in the majority of these nature, wildlife and ecodocumentaries; images that put 'nature' on display (and seldom its counterpart, the city) in a way that partly relocates the viewer to the space of the 'Cinema of Attractions', Tom Gunning's idea of early film culture where the images in themselves were more important than the story (2007: 13–20). But it can also be connected to the exotic documentary film (Bordwell and Thompson 2003: 184–185) in the tradition of *Nanook of the North* (1922) all the way up to the notorious Italian *Mondo Cane* (1962), where the images of nature are exclusively interpreted through a Western world view. And thanks to the inherent cultural veracity of the documentary, these effects usually stay transparent.

Two ecodocumentaries that highlight these problems, at the same time as they try to resist them, are *The Dark Side of Chocolate* (2010) and *Babies* [*Bébé(s)*, 2010] both centring on children as a global concern, but also as a transnational wonder of nature. Both take part in and use a broad selection of international locations for the imagery, which of course creates an aura of transnationality, while it also evokes three concerns central to ecocinema: (1) with all this travelling these documentaries have been very expensive to produce, (2) the carbon footprint left by the travelling challenges the status of these films as ecodocumentaries, (3) the combination of 1 and 2 shows that these documentaries, and most transnational ecodocumentaries alike, are an affair for the wealthy parts of the world, who also, of course, create most of the environmental problems that exist in the world. And no matter how you may twist and turn this situation, it's always going to be a biased one. Among others things because of the fact that the hegemonic privilege to formulate and articulate the agenda or 'the problem'—and therefore the 'reality'—is almost solely located within the confines of Western media (Gustafsson 1989: 22–45). Hence, to create ecological awareness and to educate the public, according to Bryan Norton's notion of environmentalism, the ecological issue must be seen from a more synoptic and contextual perspective (1991: xi), or in this case a transnational perspective.

The Dark Side of Chocolate is a television documentary directed by the Danish journalist Miki Mistrati and the American photographer, director, and human rights activist U Roberto Romano. The film shows a team of journalists investigating allegations of human trafficking and child labour in connection to the cacao industry in the African states Mali and Côte d'Ivoire (which the filmmakers constantly refer to as the Ivory Coast despite the fact that Côte d'Ivoire has been the official English name since 1985). The production of

The Dark Side of Chocolate is truly transnational with a number of production companies and countries involved: DR2 (Denmark), NDR (Germany), Danida (Danish International Development Agency that supports communication for development and democracy with ties to UNESCO), MEDIA (the EU support programme for the European audiovisual industry), TSR (Switzerland), SVT (Sweden), YLE (Finland) and ERR (Estonia). In addition to its transnational production history, the locations make this documentary geographically transnational as it is shaped as a travelogue that starts out at a chocolate industry convention in Cologne and then ventures to the 'dark' continent of Africa with stops in Sikasso in southern Mali, Abidjan, the capital city of Côte d'Ivoire, ILO (International Labour Organization) in Geneva, and the Nestle headquarters in Vevey.

With large international chocolate manufacturers such as Cargill, Archer Daniels Midland Mars, Hershey's, Nestle, Guylian, Barry Callebaut and Saf-Cacao, one would think that the focus of the investigation would be on these transnational corporations and their, perhaps, scrupulous business practices. Instead, after a short introduction to the subject, the main focus is on the African states where Miki Mistrati goes undercover and uses hidden cameras to disclose the illegal trafficking and child labour in the cacao plantations. The main reason for this division is the fact that the chocolate industry (other than issuing a statement) (CNN 2012) refused to cooperate with the filmmakers. Instead of trying to probe deeper into these global corporations the filmmakers chose to travel to Africa, which turns the film's classification as an ecodocumentary problematic due to the simplistic connotations of 'nature' that 'Africa', by this definition, signals. (see, for example, Eco Film Club 2012). The changing focus also shifts the blame to the African side of the situation, which was made possible since this is done without any sufficient contextualization of the African situation but, instead, is based on colonial and neocolonial notions and images of 'Africa'. The filmmakers do not, for example, mention the coup d'état in 1999 and the two following civil wars in Côte d'Ivoire that have influenced the whole situation. Instead they conduct a number of interviews with locals and a few officials without any real power. They also have a hard time finding the 'smoking gun', that is, actual child labourers at work in a cacao plantation, and when they do find them, the footage is obviously staged in that it is the filmmakers who tell the children to act in front of the camera. When they locate a young girl that has been abducted, it soon becomes clear that this abduction was made with the consent of her parents, who 'will be angry' when she comes home without any money. Thanks to 'our' strong Eurocentric notion of Africa as a backward and yet 'natural' continent these images connote 'truth' despite the fact that cultural (and economic) differences are only transferred through a one-way lens.

This perspective strengthens the notion that the international conventions on child labour are both relative and a Western construction, and hides the historical fact that the view of children and child labour has changed radically in Europe and the United States during the last 100 years, where children, according to the sociologist Viviana Zelizer, were seen as 'useless' if they did not work (Zelizer 1985: 7–19, 56–112). The whole composition of this ecodocumentary is therefore neocolonial and conceals, for example, the condition that if the international chocolate manufacturers (and their Western customers) would pay a more reasonable amount of money for the cacao, this problem would probably not exist. The climax of the film, where Mistrati self-righteously, and in a Michael Moore-esque style, puts up a giant monitor outside the Nestle headquarters and screens *The Dark Side of Chocolate*, is supposed to show indignation. However, it only serves to displace the fact that an actual investigation has never taken place. And of course, this also focuses on Mistrati as a 'male *ecohero*' in a manner that has become characteristic of the ecodocumentary of the new millennia, which is something that Tommy Gustafsson discusses in his chapter in this collection.

On the other hand, we have the ecodocumentary *Babies,* which allows its audience to follow four babies from around the world for a year. While the production is European with the French director Thomas Balmès and French production companies Canal +, Chez Vam and Studio Canal, the locations are transnational since the babies are born and grow up in four different countries: two rich (Japan and the United States) and two poorer (Namibia and Mongolia). However, the Eurocentric perspective is in part thwarted by the fact that there are no subtitles or narration to filter down the images to the viewers. This simple narration strategy, together with the fact that the camera is only interested in the babies, has an affect on the transnational implications, and hence the ecomessage, of the film.

By paralleling the four 'stories' the similarities between the babies become striking, regardless of the parents (who are more or less visible) and different milieus and child-rearing social habits. Nonetheless, these cultural and economic factors have the audiences in a tight grip, which is, for example, something that Peter Hartlaub expresses in the *San Francisco Chronicle,* in his review of *Babies,* where he repeatedly uses words such as 'shock' and 'shockingly' to describe his reaction to cultural clashes related to child-rearing and the different levels of consumerism—before he notes that the 'tiny people are pretty much the same' (Hartlaub 2010). *Babies* does not just demonstrate striking similarities between people on a transnational level—or rather infants, which can be seen as a kind of human tabula rasa—but, simultaneously, also works as an ecological distorting mirror, which challenges the culture of consumerism in the wealthier parts of the world: do we really need all these products and articles to survive or to gain a better life? As Roger Ebert notes, the Japanese and American children are 'surrounded by a baffling array of devices to entertain them' while the African child is equally content with a stick that is not 'made of plastic and ornamented with Disney creatures' (Ebert 2010).

Hartlaub does, however, question the film's use of two 'rich' examples: 'the urban babies in Japan and America have similar enough home lives that they start to seem redundant', and that it would have been more interesting if the filmmakers had found another baby to replace one of the rich ones (Hartlaub 2010). Nevertheless, Hartlaub does not question the two 'poorer' examples, thereby equating them as more exotic and foreign than the rich examples. Despite the fact that *Babies* uses unusual measures to avoid these types of interpretations, Hartlaub's words show, somewhat ironically, that consumerism seems to be more natural to our understanding of the (Western) world than 'nature' in a wide sense is. This also explains why one of the two 'rich' examples appears redundant since they cancel each other out, and this, in turn, would seem to suggest that these documentaries are not aimed at the poorer parts of the world.

On the other hand, the film's narrative tactics creates an impregnable greenwashing bubble that actually isolates these four cultures from each other. The almost singular focus on the babies and their immediate surroundings excludes other types of social realities; most notably those of gender and ethnicity, and therefore present them as 'natural' in a dubious way. One scene in particular comes to mind. Little Namibian girl, Ponijao, discovers that there are in fact different sexes as she, first, examines the penis of a male baby sibling (which is something that seems to be acceptable in the film's context) and then by examines her own genitals in front of her older brother. The latter, brusquely, makes her stop, which, of course, is nothing other than a blatant display of patriarchal power.

Scratch the surface and these types of normative social constructions show up everywhere in *Babies*, but they are allowed to pass by without comment and are, as a consequence, presented as 'natural' and even as transnationally shared elements to child-rearing and the social construction of gender. Considering that the film-maker most probably had hundreds, if not thousands, of specific instances to choose from, choices like this point to the imposition of a moral framework to this transnational equation.

When analysed from a transnational perspective, both *The Dark Side of Chocolate* and *Babies* end up as neocolonial yardsticks for how ecological films should look and be, while any sense of ecocinecriticism is affected by questionable audiovisual constructions of gender, ethnicity, nationality, and not least, nature. But does this mean that these films are, or should be, disqualified as ecofilms? The answer to that intricate question is no. The main reason for this is that these films are still important as political ecodocumentaries with a potential to influence, raise awareness, and take part in dialogue about the implications of the films' specific themes, although the spectators should be aware of the imposed and constructed meanings inherent to both the audiovisual and social worlds of the films. While we may not agree with their ecological or social messages, it is not up to us to disregard these (sincere) ecological efforts. Instead, we need to widen the depth of our ecocritical scope by including additional, sometimes opposing, cinematic and ecological perspectives in dialogue; in fact, it is such modes of dialogue, precisely, that will pave the way for a more comprehensively transnational understanding of both cultural production and its societal scope. The transnational aspect of the ecodocumentary (and ecocinema in general) does not become ecocritical just because of the actual crossing of national borders. Instead we propose that combining the transnational and the ecocritical becomes particularly useful due to both paradigms' openness to diversity and difference. To be ecocritical does not mean restricting one's scope of analysis to only those perspectives with which one agrees. To operate on a transnational scale necessitates expanding one's scope beyond simplistic conceptualizations of nations and unidirectional flows of culture and commerce. To achieve a point where transnational ecocinema emerges as a useful interrogative synergy, one needs to maintain awareness of the contradictory and multidirectional forms of rhetoric of which contemporary film culture consists.

Several chapters in this collection also address the transnational ecodocumentary in a critical and innovative way. For example, Enoch Yee-Lok Tam discusses Chinese television documentaries and poses the vital question: to what extent do ecodocumentaries have a greenwashing effect, aimed both at the national and the international community attempting to come to terms with China's growing industrialism and environmental problems that transcend its national borders. From another angle, Chu Kiu Wai explores the way in which transnational circulation of commercial products and cultural materials facilitates both global economic and ecological depravation, that is, a situation where the global interconnectedness and respect for nature seems no more than ecocritical rhetoric. And finally, Ilda de Teresa de Castro explores the anthropocentric and ethic relationship between humans and non-humans in the ecodocumentary, suggesting that the critical vantage point afforded by a transnational approach can foster new perspectives on human interactions with their animal 'others'. All of these angles on the ecodocumentary 'genre' emphasize the great variety of approaches that ecocritics must consider when exploring the 'factual' content of ecological media.

Transnational Fictional Feature Films

To complement these 'factual' variations of transnational ecocinema, we must also engage with fictional texts as their claims to political participation operate from a somewhat different perspective. Many chapters in the collection focus on fictional films, some on the more mainstream cases emanating from Hollywood (Gustafsson), some on blockbusters from China (Corrida Neri's discussion of Feng Xiaogang's films). Even in these mainstream commercial films, ecocritical and environmentalist concerns emerge in a range of ways. They utilize the environment for consumerist or ideopolitical purposes, while they also participate in public discussion in ways that claim to increase ecological awareness. While cynical questions over producer motivation are certainly valid here as well, would this invalidate the films' aspirations to making a connection with audiences and increase their knowledge of ecological concerns? After all, even greenwashing perspectives—approaches which use environmental or sustainable rhetoric to justify insustainable or exploitative means and goals—may generate discussion on the ecological connotations of cinema. Other chapters take a more explicitly art house–based view of ecocinema. Pietari Kääpä's overview of transnational considerations in ecocinema, for example, focuses on the films of Jim Jarmusch and Gaspar Noé, both of whom challenge mainstream conventions by providing subversive impressions of their ideological connotations. The integration of substantial ecophilosophical and ecocritical material into both filmmakers' subversion structures makes them particularly interesting case studies for emphasizing the diversity of ways in which filmmakers envision the relationship between the environment and humanity's place in the ecosystem.

Transnational concerns underline all the chapters focused on fictional films as they come to indicate the ways in which cinematic depictions of ecological considerations are rarely contained within human-made borders. To explore the implications of fictional films for our conceptualisation of transnational ecocinema further, we discuss one of the most debated examples of transnational film culture, Stephen Frears' *Dirty Pretty Things* (2002). Many scholars have discussed this film from a transnational angle (Ezra and Rowden 2006; Loshitzky 2010; Van de Ven 2009; Prime 2006) while others discuss it from an ecocritical perspective (Stein 2010). While Stein's reading of the film as a critique of the exploitation of immigrant bodies in transnational Europe is an entirely appropriate interpretation, we suggest that this sort of critique can be taken further by not only seeing the film as a reflection and reaction to injustice, but also as part of the wider problem of systematic exploitation. What we have in mind is to propose an alternative reading of the film that varies from the more 'canonic' takes that frequently circulate in transnational academic film culture. This is not to imply that our suggestion that the film exhibits certain neocolonialist aspects should be seen as the only or even the preferred reading. Instead, this critical take on a film that we acknowledge makes a valuable contribution to global cultural politics is an indication of the need for polyvocality in thinking about ecocinema. Indeed, when discussing film culture, ecocritics can be too hasty in approving of any film with a green agenda as an immediate force for good. To achieve a sense of critical weight, ecocinema studies will need to be explicitly critical of its own parameters, as well as those of other fields, in order to be able to make sufficiently dynamic advances towards the integration of the humanities and the sciences. To attain this critical perspective, we must be willing to criticize films we otherwise find agreeable and use any findings to rethink the ever-expanding critical body on ecocinema. Thus, the necessity to be open to readings that go against the grain of 'common' ecocritical sense is a necessary approach for a thriving academic analytical practice.

Organ Harvesting in Culture and Cinema

The central theme of *Dirty Pretty Things* concerns the harvesting of organs from vulnerable immigrants and the political economy that benefits from these exploitative practices. Organ harvesting has a long history in arts and cinema, often acting as a source of revulsion or a symptom of madness in a range of horror films from *Frankenstein* (1931) to *Les yeux sans visage/Eyes Without a Face* (1960). The transplant has been given agency through a set of traces remaining from its original body, which then provide the topic for the narrative's inherent conflict—these range from a writer inheriting the organs of a murderer in *Body Parts* (1991) to, more problematically, a heart transplant from a black lawyer into a racist cop in *Heart Condition* (1990). While the relationship between the implant and the host body is depicted in these genre films, the majority of these films ignore cultural and biopolitical concerns in favour of narratives that play on fears of losing one's autonomy.

If they do show cultural differences, these are in comedic mode and enforce a range of stereotypes.

Depictions of organ harvesting are similarly stereotypical in mainstream representations, with images of stolen kidneys proliferating in films like *Jay and Silent Bob Strike Back* (2000). Perhaps the most explicit and intriguing of these sorts of narratives is the survivalist horror tale *Turistas* (2005), a film focused on a group of American tourists who get lost in the jungles of Brazil, only to be hunted by a local organ dealer. The film shows some initial awareness of the complexities of Western organ tourism as local inhabitants refuse to be photographed with the tourists for fear that their children are being earmarked for organ speculators. As tables are turned on the tourists, a satirical portrayal of the rise of the exploited momentarily emerges. Soon, the film turns into something very different as the protagonists fight back against the very clearly insane villains out to harvest them. Any sort of reactionary critique of Western exploitative practices and awareness of transplant tourism and American conceptions of its others as resources is realigned as the survivalist narrative re-establishes existing relations of domination. Medical practitioners who refuse the normative directions of transplantation biopower are shown as maniacal 'liberators of the people', not unlike the Latin American 'dictator' image perpetuated by conservative American media. In this sort of mainstream cinema, we are never afforded the opportunity to explore the socioeconomic circumstances of non-white donors in full nor are we afforded any chance to consider non-white hosts of organs. The picture that emerges is only about white recipients and non-white bodies as corporeal resources, where any deviation from this positioning will result in corrective action.

If the transplanted organ acts as a concrete and metaphoric site of biopolitical inequality and, ultimately, the reassertion of dominant values in mainstream cinema, an obvious way to subvert this unidirectional trade would be to discuss films that are produced outside the system. Such films would presumably be produced by non-Hollywood industries or independent filmmakers and circulate the world (if they do) via festivals and specialist distribution. Based on its theme, *Dirty Pretty Things* fits the bill as an alternative to the mainstream. It tackles the problems of organ harvesting directly and provides an examination of its exploitative practices from the 'other' perspective. Its depiction of the biopolitical economy of transplantations focuses on a Nigerian doctor Okwe (Chiwetel Ejifor) and a Turkish immigrant Seyna (Audrey Tautou), who work as the night porter and cleaner at the Baltic Hotel. Senor Juan (Serge Lopez) runs an illegal organ harvesting operation from the hotel and tries to blackmail Okwe to work for him. Okwe is able to turn the tables on Serge with the help of the prostitute Juliet (Sophie Okenado). The film culminates with Serge's kidney providing funds for Okwe to return home under a new identity and Senay to fly to New York. This culmination is certainly significant as it provides a sense of 'resourceful resistance' to the typical flow of the organ trade and establishes a multicultural form of community action against oppression (Stein 2010: 110). Simultaneously, the film shows a keen sense of awareness that there may be no happy endings as both protagonists face uncertainty and liminality in their new home contexts.

In comparison to genre fare like *Turistas*, *Dirty Pretty Things* is taken very seriously. It is critically respected, has won many prestigious awards, generated political discussion in the media and attracted the attention of academics. While the film is potentially the most advanced of the organ harvesting films and close to reality in capturing the oppressiveness of the social milieu that fosters this exploitation, how solid are its claims to be 'serious cinema'? The film was produced for the BBC by Calador Productions as their first production venture and distributed by Miramax in the United States. Both BBC and Miramax are associated with a certain level of prestige, and are especially well known for their work in the arts and politically oriented media production. Yet the productions they frequently support are more appropriately seen in the category of the 'popular art film' (Cook 2010) rather than as some sort of alternative, truly esoteric fare. Indeed, it is this sort of cultural-industrial liminality (or self-conscious use of both populist and artistic conventions) that characterizes *Dirty Pretty Things*. For one, its characters are not much different from the prototypes we see in genre cinema. The heroic male Okwe must protect the fragile Senay from the dangers awaiting in the Baltic while the resourceful community of immigrants provides support for his revolution against the oppressively sleazy Juan. This does not provide us any viable alternatives to the usual setup of geopolitically or culturally delineated categories of the exploiters and the exploited, of those who belong in Fortress Europe, and those who will remain external to it, even within its borders. While the film is certainly 'artistic' in its use of gritty locations and aesthetics, it also works squarely in the thriller genre. In fact, this is the marketing angle adopted for the film on its US release with posters featuring a half-naked image of its star Audrey Tautou brandishing quotes about its thrilling narrative. The use of English as the only language of communication is again a reflection of the context in which the film is set, but it also identifies the film's target audiences as ones predominantly from English-speaking contexts, even though it will also clearly travel to global audiences in different languages. Finally, the film's conclusion is certainly an instance of collective resistance, but it also trivializes the situation of harvested immigrants by emphatically underlining an outraged, reactionary sense of social injustice as Okwe pointedly states: 'We are the people you do not see. We are the ones who drive your cabs. We clean your rooms. And suck your cocks'. While the film culminates with uncertainty as to the futures of Okwe and Senay, and a lingering sense that a borderless world is only such for those at the upper echelons of society, inflammatory statements and an implied romantic conclusion seem a concession to audiences who are used to such emotional crescendos.

How does the cinematic language's close resemblance to hegemonic vernacular impact its postcolonialist critique? Elizabeth Ezra and Terry Rowden (2010) have discussed Okwe, Senay and the other immigrants of the film as 'postcolonial transplants'. Instead of discussing organ transplants concretely the transplant in question here is the immigrant's body that has been subjected to the colonizing imposition of the host society, which subjugates them as cheap human resources. For Ezra and Rowden, the immigrant body operates as the site of objectification to globalization's 'dynamics of acceptance and rejection, in which assimilation on the one hand and the maintenance of cultural distinctiveness on the other, are the agonistic poles of globalization as a lived experience' (2010: 212). Instead of speaking of transplants in this more metaphorical sense, we take the term literally as an explicitly biopolitical focus on transplantations and organ commerce, as part of the multiculturalist politics of assimilation and rejection.

The film's thematic criticism and its cinematic limitations position it as precisely the type of conflicted multiculturalist object that Slavoj Žižek's critique of multiculturalism identifies. According to him, multiculturalism as a form of sociopolitical violence works

> by incorporating a series of crucial motifs and aspirations of the oppressed—truth is on the side of the suffering and humiliated, power corrupts, and so on—and rearticulating them in such a way that they became compatible with the existing relations of domination.
>
> (Žižek 1997)

Multiculturalist rhetoric thus emerges as a way to express a sense of concern and awareness of ongoing problems in a way that evades providing any real solutions to them and maintaining the status quo. While we are shown the injustice suffered by the 'other', in *Dirty Pretty Things* the systems of exploitation are just slapped on the wrist. For one, the film culminates with the white British representative of the harvesting operations receiving Senor Juan's kidney from Okwe. The system thus stays intact with only the second-tier Spanish exploiter/exploited Senor Juan suffering for the larger systems of exploitation. The film's emotional appeals for equality and respect are granted by the collective uprising, but wider structures of exploitation remain in place—this is tolerance of the problematic kind, where a certain degree of permissiveness and discontent is allowed. But this can only be done in such a way as to pose no real systemic challenges to the hegemonic neoliberalist order, as is the case with sacrificing Juan for Okwe and Senay's escape off the grid. Instead of providing a fundamental challenge to the existing order, the escape is 'tolerated' as it poses no real threat to the larger neoliberalist structures. It is intolerance of the neocolonialist kind.

Ecocosmopolitan Analysis

To address some of the questions that this multiculturalist critique of *Dirty Pretty Things* raises, Ursula Heise's ecocosmopolitanism (2008) provides us with an alternative angle on transnational exploitation. She suggests that 'environmentalist advocacies of place assume that individuals' existential encounters with nature and engagements with intimately known local places can be recuperated intact from the distortions of modernization' (2008: 11). Responding to a pervasive tendency of ecocriticism to see causes and solutions to environmental problems as either local or global issues, she proposes the necessity to acknowledge the tangibility of environmental problems by addressing local concerns but, simultaneously, these problems need to be seen in the wider context of the planetary ecosystem. Indeed,

> most networks of information open the local out into a network of ecological links that span a region, a continent, or the world [and which] allow individuals to think beyond the boundaries of their own cultures, ethnicities or nations to a range of other sociocultural frameworks [providing] an attempt to envision individuals and groups as part of planetary 'imagined communities' of human and non-human kinds. (11)

Heise's suggestions are important as a rhetorical redirection and connect closely with our preference to see ecological problems on a transnational scale. By insisting on a continuous dialogue between the global and the local, a more pervasive oscillation between the tangibility of local space and borderless belonging, as part of the planetary ecology, expands the parameters of ecocinecriticism. To explain how an ecocosmopolitan perspective would work in practice, she uses the example of Google Earth, an application that allows users access to a global scale of spaces, all from the comfort of their own local user position. While such applications are vital for visualizing the planetary interconnectivedness of local concerns, and the ways their causes and effects operate beyond spatial specificity, the application is appropriately and ironically symbolic of the wider problems ingrained as part of the DNA of such globalized perspectives. In addition to the obvious problem of supporting data harvesting by a singular corporate entity, other issues, including the inequality of access to networked computers and their material discarding in cheap industrial conditions, often have harmful or marginalizing effects on deprived local conditions (Cubitt 2009). While it is entirely necessary to complicate the simplistic binary of the global and the local, the planetary comes with its own set of associated problems.

Ecocosmopolitanism shares similarities with multiculturalism in its emphasis on operating above and beyond borders, and in its awareness of the complexities of global imbalances of power. But it also shares multiculturalism's fallacies in that a planetary perspective may not be the most incisive way to pay due attention to material scarcity and the inability of marginalized individuals to attain the transcendent perspective of the cosmopolitan. This is made more evident when we consider the biopolitics of *Dirty Pretty Things*. While the film is of substantial merit in raising awareness of exploitative practices and global inequality, its inflammatory tone and narrative conformity reminds us of the ways ecocosmopolitanism and its sense of the planet foreground the big picture in favour of the less spectacular or sensationalist aspects of organ transplantation. It is, of course, entirely necessary to consider local ecological problems within a wider planetary scope, just as it is necessary to critique the organ harvesting 'system' by showing us its status quo, as shown in *Dirty Pretty Things*. Simultaneously, it would be similarly important to show organ transplantation experiences from a perspective other than the white one—all experiences are only based on this one way system, where the host body is connoted to be invariably white with no consideration for other types of recipients. For some, this abidance with the rules of the system amounts to a realistic approach to depicting prevailing global inequality and exploitation. For others, *Dirty Pretty Things* works by raising awareness of important issues without proposing any fundamental challenges to the prevailing norms of cinematic biopolitics.

As part of the Eurocentric notion of 'repressive tolerance'—the tolerance of the other in its aseptic, benign form, as Žižek would have it—the conclusion of *Dirty Pretty Things* only subverts the flow of organs momentarily and confirms the subsumation of biomaterial resources into the constitution of the multicultural body of violence. While conceptualizing the film as biopolitical ecocinema enables us to pay attention to the ways it critiques an existing problem of the contemporary neocolonialist order, an even more pervasive critique emerges when we contextualize this with the problems associated with multiculturalist ideology—a concern central to studies of transnational cinemas. For a truly critical take on the multicultural biopolitics of organ harvesting, we can learn from Will Higbee and Song-Hwee Lim's proposal of a critical transnationalism that encourages us to 'interpret more productively the interface between global and local, national and transnational, as well as move away' from the Eurocentric tendencies that may prevail in cinema (Higbee and Lim 2010: 10). Conceptual ambiguities, of course, permeate this definition as well, but to us, it seems more productive than the explicitly expansive ecocosmopolitan perspective. For Higbee and Lim, this approach proposes 'critical understanding of the political imbalances as well as the unstable and shifting identification between host/home, individual/community, global/local, national/transnational, as well as the tensions these generate' (Higbee and Lim 2010: 12). This implies that the analyst cannot take for granted that a 'mere' transnational approach is able to provide the necessary critical distance— in fact, observed within the framework of conventional transnational film studies, *Dirty Pretty Things* appears as an entirely comprehensive critical depiction of biopolitical exploitation.

By shifting focus, firstly, to critical transnationalism's call to interrogate the principles that underlie transnational film studies, and secondly, combining them with matters concerning ecocinema—namely, biopolitics—we can get a better perspective on areas *Dirty Pretty Things* does not address. Moving from the conflicts between immigrants and host societies to an explicit focus on organs as both commodities and culturally inflected biomaterial enables us to consider the ways multiculturalism-as-violence operates. We are aware that the film can be interpreted on a metaphoric level where organs—the postcolonial transplants—act as signifiers of exploitation of immigrant bodies. But this metaphoricism also enforces the sense that we are not seeing the whole picture. While individuals may choose to assimilate or reject multicultural societies, their harvested parts have no such ability to contest. If we consider the grafting of the body part to its new host body, immunosuppressants and a range of technologically advanced forms of operation make sure to erase any sort of contentious trace of rejection from the allograft. While the new 'body politic' is a physically heterogeneous body, its fully functional form is one where differences should not exist. This is not unlike the multiculturalism critiqued by Žižek, which makes it pertinent that we consider the connotations of *Dirty Pretty Things* beyond what we are shown in the film— namely the ways organs act as transnational biomaterial once they are removed from the (involuntary) donor.

Ecocinema's Biopolitics: The Administration of the Human Body

A crucial part in the construction of this homogeneous body of cultural representation is the lack of available alternative perspectives of, for example, organ hosts from underprivileged contexts. Similarly, dominant perceptions circulating in mainstream film culture underline ethnic differences in the donor/host relationship but also fail to consider the underlying implications of structural differences in representing allografting. An example of this can be seen in Nick Cassavetes' *John Q* (2001), which touches on class and race issues through its focus on a black proletariat family, whose son faces death due to his father not possessing adequate medical insurance coverage, but trivializes its critical potential by relying on a range of caricatures and unsubtle rhetoric, such as the use of Ave Maria on the soundtrack accompanying the demise of the angelic—and in this case, problematically—white donor (see Heinz 2009, for more discussion of some of these fallacies). The lack of alternative perspectives connects to wider social realities as Goode et al. (2007), for example, argue that medicinal, racial and ethnic disparities affect transplantation outcomes. These are not so much to do with genetics but societal factors such as lifestyle, income, environmental conditions, and means of sustenance. Such studies suggest that this is a systemic problem, where the success or failure of an allograft may indeed be ascribed to biological genetic factors, but we must ultimately address underlying socioeconomic issues and prevailing prejudiced attitudes about transplants if we are to truly understand immunosuppressant reactions and failures as systemic concerns.

While *Dirty Pretty Things*' focus on immigrant experiences is heavily critical of the current neoliberal order, dominant patterns of host and donor bodies remains unchallenged. By avoiding questions to do with unsuccessful allografts or showing the lack of infrastructural support to advance them, cultural representations work like anesthesia, not unlike the immunosuppressants that overcome biological differences. In the scale of global film culture, it complements the prevalent conceptions of organ harvesting in the contemporary biopolitical order and strengthens the global perception of the other as a source of exploitation. Nor does this approach provide or propose any real practical solutions to these problems of difference in biopolitical status.

To avoid reaching an unproductive rhetorical cul-de-sac, we must emphasize that *Dirty Pretty Things* remains an important film in its cultural biopolitics. But similarly, it is clear that its content enables readings of an entirely different variety, even ones that see it as only contributing to neocolonialist suppression. These rhetorical fluctuations raise a number of questions related to how film producers working in an inherently commercial form of industrial production can ever hope to overcome the limitations imposed on them by the wider industrial and cultural structures in which they operate. Can films truly aspire to instigate change as ecocritics would wish them to do? This is a question that is debated and addressed from many angles in this collection. For now, this discussion of *Dirty Pretty Things* emphasizes the need for a further degree of criticism in ecocinema studies. Combining the calls for self-criticism and pervasive social injustice evident in critical transnationalism with the calls for social and environmental justice inherent in ecocriticism provides the key to our proposal of transnational ecocinema as a viable critical response to ecocosmopolitanism. Starting out from the paradigmatic advances of ecocosmopolitanism, this collection proposes the adoption of transnational approaches to focus and clarify some of its ambiguities and abstractions. Thus we move away from the conception of transnational films and film culture as only concerning difficult or marginal films and, instead, see it as comprising a range of approaches that share the theme of cross-border collaboration and concerns with imbalances and inequalities of power in global society. As such, transnational ecocinema is a form of critical analysis that interrogates the contradictions, complexities, obstacles and opportunities emerging in the clash of the local and the global. That it does this by committing to advancing ecological understanding in film culture and film studies allows it to address both ecosystemic concerns and the biopolitics of human ecology in ways that necessitates us to question any complacency in ecocritical or transnationalist thinking and push them in further critical directions.

References

Bordwell, D. and Thompson, K. (2003), *Film History. An Introduction*, New York and London: McGraw-Hill.

Brereton, P. (2005), *Hollywood Utopia. Ecology in Contemporary American Cinema*, Bristol: Intellect.

CNN (2012), 'Global chocolate and cocoa industry' http://i2.cdn.turner.com/cnn/2011/images/04/06/response.pdf. Accessed March 19, 2012.

Cook, P. (2010) 'Transnational utopias: Baz Luhrmann and Australian cinema', *Transnational Cinemas*, 1: 1, pp. 22–37.

Ebert, R. (2010), 'The babies are cute. Well, all babies are cute', *Chicago Sun Times*, May 5, 2010.

Eco Film Club (2012), 'The Dark Side of Chocolate' http://sadhanaforest.org/wp/2011/07/the-dark-side-of-chocolate/. Accessed March 19, 2012.

Ezra, E. and Rowden, T. (2010), 'Postcolonial transplants: cinema, diaspora and the body politic', in M. Keown, D. Murphy and J. Procter (eds), *Comparing Postcolonial Diasporas*, New York: Palgrave, pp. 211–227.

Goode, T., Isaacs, R. and Hricik, D. (2007), 'Ethnic and racial diversity in transplantation: does everyone benefit equally', *Advanced Studies in Medicine*, 7: 9, pp. 268–274.

Gunning, T. (2007), 'The cinema of attractions: early film, its spectator and the avant-garde', in P. Grainge, M. Jancovich and S. Monteith (eds), *Film Histories. An Introduction and Reader*, Toronto and Buffalo: University of Toronto Press.

Gustafsson, L. (1989), *Problemformuleringsprivilegiet*, Stockholm: Norstedts.

Hartlaub, P. (2010), 'Review: "Babies" same despite different homes', *San Francisco Chronicle*, May 7, 2010.

Heinz, J. (2009), 'Scheming on screen: the art of audience manipulation as manifested in *John Q.*', *Communications*, Spring 2009, pp. 109–115.

Higbee, W. and Lim, S. (2010), 'Concepts of transnational cinema', *Transnational Cinemas*, 1: 1, pp. 7–21.

Ingram, D. (2000) *Green Screen: Environmentalism and Hollywood Cinema*, Exeter: University of Exeter Press.

Ivakhiv, A. (2008), 'Green film criticism and its futures', *Interdisciplinary Studies in Literature and Environment*, 15: 2, pp. 1–28.

Loshitzky, Y. (2006), 'Journey of Hope to Fortress Europe', *Third Text*, 20:6, pp. 745–754.

—— (2010), *Screening Strangers: Migration and Diaspora in Contemporary European Cinema*, Bloomington: Indiana University Press.

Lu, S. H. and Mi J. (ed.) (2009), *Chinese Ecocinema in an Age of Environmental Challenge*, Hong Kong: Hong University Press.

Lykidis, A. (2009), 'Minority and immigrant representation in recent European cinema', *Spectator*, 29: 1, pp. 37–45.

Norton, B. (1991), *Towards Unity among Enviromentalists*, Oxford: Oxford University Press. Ortega, O. R. (2011), 'Surgical passports, the EU, and *Dirty Pretty Things*', *Studies in European Cinema*, 8: 1, pp. 21–30.

Prime, R. (2006), 'Stranger than fiction: genre and hybridity in the "refugee" film', *Post Script*, 25: 2. http://www.freepatentsonline.com/article/Post-Script/143723767.html. Accessed March 19, 2012.

Stein, R. (2010), 'Disposable bodies: biocolonialism in *The Constant Gardner* and *Dirty Pretty Things*', in P. Willoquet-Maricondi (ed.), *Framing the World: Explorations in Ecocriticism and Film*, Charlottesville: University of Virginia Press, pp. 101–115.

Van De Ven, K. (2009), 'Just an anonymous room': cinematic hotels and motels as pneumonic purgatories', in D. Clarke and M. Doel (eds), *Moving Pictures / Stopping Places*, Lanham: Lexington, pp. 235–255.

Zelizer, V. A. (1985), *Priceing the Priceless Child. The Changing Social Value of Children*, New York: Basic Books.

Žižek, S. (1997), 'Multiculturalism or the cultural logic of multinational capitalism', *New Left Review*, 225, September–October, pp. 44–45.

Transnational Approaches to Ecocinema: Charting an Expansive Field

Pietari Kääpä

Environmental engagement has become the new topic *du jour* in film studies. We have seen an increasing amount of popular and art house films from different cultural contexts engage with ecological issues. Academic works such as Sean Cubitt's *Ecomedia* (2005) and David Ingram's *Green Screen* (2000) have made significant contributions to expanding cultural ecocriticism from what was largely a literature-focused field to one that interrogates the relationship between media and ecology. Other works, such as Pat Brereton's *Hollywood Utopia* (2005) and Robin Murray and Joseph Heumann's *Ecology and Popular Film* (2009), have responded to Cubitt and Ingram's calls by tackling a range of individual films that signal (at least on the surface) the increasing 'greening' of Hollywood, exploring both positive (increasing ecoawareness) and negative (the possibility of 'greenwashing' environmental issues) aspects of this cultural industry. In this wide-ranging critical discourse, theoreticians and film historians ponder ways for cinema to take part in the political debates over environmental issues and sustainable practices. Yet, much of this work has not progressed beyond criticism of large-scale Hollywood blockbusters, which are often content with presenting populist arguments about global warming and other environmental problems. While it is crucial to continue the debate instigated by the above authors on mainstream popular cinema, as this ultimately enables reaching as wide audiences as possible, the ecopolitics of contexts other than Hollywood have received substantially less attention.

Recently, Sheldon Lu and Jiayan Mi's *Chinese Ecocinema* (2009) has explored these concerns in the cultural and sociopolitical context of global China, interrogating the ways cultural producers contribute to political debates on the planetary implications of China's economic transformations. Their intervention is of substantial importance as it presents the first large-scale attempt to move beyond Hollywood dominance of the field by providing a comprehensive and systematic study of ecological concerns in the different cinemas that make up the wider body of 'Chinese cinema' (mainland, Taiwan, Hong Kong). Despite these advances in exploring the planetary implications of both Hollywood and specific national cinemas, more specific interrogation of the complex interdependent networks of the global ecosystem needs to be conducted. While the existent paradigmatic division between Hollywood and 'national' cinemas is certainly understandable, the division also corresponds closely to a prevalent limitation of ecocritical writing observed by ecophilosopher Ursula Heise (2008). According to her, much of the work in the field is constructed along a binary paradigm between planetary and local concerns. Here, writers work from the basis of an either/or approach, choosing to espouse the 'authentic' virtues of localism or the ecosystemic holism of planetarist rhetoric. The academic focus on either global or national concerns in ecocinema is premised on a similar set of assumptions, where the planetary/universal scope of global Hollywood contrasts with the more specific, place centric concerns of the local/ national rhetoric. From Heise's perspective, the local is too idyllic(ized) and restricted in scope to account for the ways environmental problems impact the wider ecosystem. Meanwhile, the utopianism and non-specificity of the global disables it as a way to comprehensively explore ecological issues or to account for environmental problems.

Following from this acknowledgment of the limitations of ecocritical analysis, this chapter has two analytical reorientations in mind: (1) proposing alternative modes of ecocinematic analysis beyond the dominant binary of Hollywood/national cinema; (2) laying foundations for analysing ecocinema from a range of transnational angles including content, production, distribution, exhibition and reception. As the case studies of the chapter move from Spain to Tokyo, from the Three Gorges in mainland China to Hong Kong, the oscillation between the specificities of place and the cultural political transcendence of ecosystemic concerns emerge as key topics of the type of transnational ecologicalism (exploring the transnational dimensions of ecological thinking) this chapter seeks to encourage. The conclusion will expand on this argument to provide some suggestive directions for discussing the materialist-industrialist connotations of ecocinema and the ways reception studies can contribute to and ultimately enhance our understanding of the sociopolitical potential of films with an environmental conscience.

Hollywood Cinema's Ecocatharsis

To understand how synergizing ecoritical thinking and transnational film studies can advance both fields, we must start out by interrogating how the ecological work of Hollywood cinema can be seen in globalist ways. It would certainly not be overstating the case to suggest that Hollywood films function as the dominant arbiter of audience perceptions on ecological thinking. Much of contemporary Hollywood cinema conceives of nature as the property of humanity, relying on anthropocentric conceptualizations that colonize the environment for human consumption (see the works of Ingram and Brereton for more on this). Of course, humanity's role in the ecosystem should not be ignored nor should we argue that the ecosystem is only devoted to human reproduction and sustainability. Yet, this chapter is not concerned with ecophilosophical arguments over humanity's place on earth—for us, it is a matter of fact that humanity plays an indelible reciprocal role in the planetary ecosystem.

Instead, it is more urgent to interrogate the different ways ecocritical awareness manifests in global cultural politics of representation, and the materialist-industrial concerns that affect and underline these developments.

Part of this reorientation is a pointed engagement with increasingly powerful and widespread conceptualizations of the global environment, such as those emphasized by commentators like Thomas Friedman (2007). For him, globalization involves an inevitable deterministic process that will 'flatten' the world through its enabling potential for developing economies such as those of China. While he suggests that globalization results in the dissolution of global power distances and economic inequalities, his focus is inherently US-centric and protectionist in its rhetorical stance. This sort of self-critical, but self-serving rhetoric also extends to ecological argumentation in much of Hollywood's ecocritical film production. Greenwashing of ecocritical concerns through visual spectacle is already present in early cinema such as *Oil Wells of Baku: Close View* (Lumiere 1896), which works simultaneously as 'an ecological film, an environmental film, and a history of wealth. It also foregrounds a history of spectacle and the history of one of the most contentious of modern currencies' (Murray and Heumann 2009: 35). Crucial in this view is that such films can occupy multiple positions simultaneously—they can be critical reflections on historical explorations of industrial activity and sustainable practices, while they simultaneously act as celebrations of wealth and spectacle. In films such as *The Day After Tomorrow* (Emmerich 2004) and *2012* (Emmerich 2009), ecospectacle provides an easy access point for spectators interested in green politics or even those who feel 'obligated' to participate in environmental activity either out of moral awareness or peer pressure. Here, the emotional highs and lows of a blockbuster narrative allow the spectator to live through the ecocatastrophe in as entertaining and easily-digestible ways as possible. This provides a safety-net for avoiding asking real questions about environmental degradation and the role of humanity in not only past disasters, but in the current and ongoing anti-environmentalist activity. Instead, the potential to experience the trauma of impending ecocatastrophe creates a sense of ecocatharsis, which, in a sort of cinematic hypnotherapy, relies on the inclusion of geopolitical ironies such as the illegal migration of US citizens to Mexico in *The Day After Tomorrow* or the White House being crushed by the USS John F. Kennedy air force carrier in *2012*.

These instances displace real concerns into largely inconsequential geopolitical humour, underlining the irony of blockbusting ecocinema (greenbusters) in that these films often try to make a point of shattering the 'false consciousness' by which the American culture industry naturalizes its global cultural dominance. Both *The Day After Tomorrow* and *2012* include substantial criticism of the failures and short-sighted measures of the US-led international political system. Yet, these instances of criticism are conducted within an implicit and unmentioned assumption of the natural leadership of Western, pro-US democracy. This notion comes through especially in the martyrdom allowed for the morally awakened leaders of the United States, which is, of course, not entirely surprising considering the industrial-political roots of the films. Implicit in such argumentation is the suggestion that working with capitalism requires trust in its systems of exploitation. This uncomfortable reconnection extends the Fukuyamaian ideological end of history to ecological thinking as the films are entirely hesitant to do away with old political-cultural structures, a notion made especially clear as the ultimate signifier of neo-conservative American-centric authenticity of the nuclear family is restored in both cases.

The disasters predominantly function as processes of purification as they work to get rid of some of the excesses of the contemporary geopolitical system. While a pristine new world governed by enlightened, leftist individuals is indeed created, existing large-scale technological order and supremacy is still in place, supported by all the hallmarks and the institutions of the antecedent era. Instead of carrying out their seemingly leftist criticism of global America, the films only strengthen neo-conservative greenwashing as the global norm for ecopolitics, while they also maintain the cultural-industrial hegemony of the Hollywood system for representing these issues. By situating themselves as the global standard through pervasive marketing and blanket coverage in the international media, these films cater for ecologically curious audiences who receive their messages within clearly constructed environmentally aware parameters. Much as with the flattening of the world, these critical reflections are not so much to do with some sense of ecocritical philanthropy. Rather, they are more appropriately considered a form of 'feel-bad' cinema, which acknowledges environmental problems and even the spectators' complicity in these problems, but where the visual pleasures of the text also allow the spectator to dismiss any 'overt' concern over the issues represented.

Let us illustrate this inactive activity through a scene from a seemingly unrelated Hollywood comedy *She's Out of My League* (Smith 2010). As Kirk (Jay Baruchel) introduces new girlfriend Molly (Alice Eve) during an awkward family dinner, the focus turns to the scatological antics of Kirk's family and their recycling habits, which include restrictions on flushing the toilet for 'number ones'. To enforce the use value of this green activity, the father cites the predominant authority on the issue of global warming—Roland Emmerich's *The Day After Tomorrow*. According to the producers of the film, global warming is something that large numbers of people—especially of an upper middle class liberal persuasion—take seriously. Yet, this green awareness is shown to be of a substantially shallow quality as it only acts as the source of scatological humour and a way of criticizing middle class complicity. Ecological issues are thus subjugated to character-based humor on liberal America, consisting largely of the target audience for Emmerich's populist green politics. While this aside should not be taken too seriously, the implications of the 'joke' can tell us of some of the ways in which ecocritical rhetoric is neutered by the vocalization of such ideas in the mainstream, urging us to reconsider the oft-made claims about the wide sociological impact of green blockbusters.

Hollywood and the Ecological Panoptic

What impact can these films achieve in their ecocriticism, especially as their 'climate science' is often very easily (and very publicly) disproven? Can they, in fact, be considered as part of the machinery that keeps the climate change skeptics going? If we consider populist environmentalism as greenwashing, we can think of this sort of circumventive discrediting of advances in climate science as operating as a cultural version of panopticism. As conceptualized by Michel Foucault (1995), panopticism involves the creation and fostering of social power by making the individual their own controlling agent, as they see the prevailing forms of social order as the expected and accepted norm and uphold these structures through their willingness to participate as a productive member of society. Through socialization, capitalism and its ethic and moral codes maintain their dominant hold and mastery over the general public in ways that are often invisible and therefore more powerful. According to Hwang Sung-Uk (Hwang 1998), mainstream ecocinema works in similar ways to panopticism as it draws on currently popular and accepted notions of green awareness—that is, the pervasive pressure from all aspects of society to think and act green. While environmental awareness, recycling and so on are vital for sustainability and the ethical conduct of an individual, a large part of this approach in mainstream ecocinema is the normalization of the rhetorical arguments and frameworks of the geopolitical and geo-economic order.

A further problem with these Hollywood films is their myopic view of the world as their panoptic perspectives extend globally. As forms of mainstream ecological entertainment, greenbusters only provide occasional and highly stereotypical glimpses of life beyond the Western hemisphere—rarely do we see anything beyond the marketable exoticism of non-Western cultures. By consistently demonstrating how other parts of the world are 'undeveloped' (or considered in these terms—see *2012* and abandonment of the Indian team of scientists), the politics of scarcity are normalized. Such depictions are doubly problematic as the films' (superficial) racial equality works as a sort of preemptive answer to any criticism for failing to portray other cultures in equal terms. The self-sacrifice of an Obamaesque president can be considered as one very insistent indicator of the film's leftist credentials, which compensate for fallacies in its claims to cultural sensitivity. This is only one aspect in which the film's cultural politics work to sideline discussion from addressing its reliance on conservative values and technological prowess only achievable through reliance on capitalist profiteering and neoliberalist propaganda. In many ways then, the industrial and cultural functions of Hollywood ecologicalism often provide a largely toothless and self-serving critique of US geopolitical power and the vices of capitalism. But these are ultimately aspects that allow such films to travel despite their implicit position on US hegemony, all which works to promote US power as the only real answer to global ecological destabilization.

Transnational Ecocinema

In contrast to these limiting normalizations of a new world order, transnational cinema studies focuses on texts that centralize the theme of global discontent and inequality. In contrast to the flattening of the panoptic rhetoric, the critical work of the field aligns with Nathan Gardels' perspective on global order: 'we may now be entering a new phase: post-globalization which is less characterized by a flattening of old difference than the appearance of new ones' (2008: 1). Environmental issues are just one very visible instance of the type of inequality globalization sows. One way of addressing these issues is to shift the focus to texts that consciously challenge and complicate our relationship with cathartic modes of spectatorship. While it is absolutely necessary to conduct place-centric contextual analyses of different ecological texts and the issues they cover (such as we see in most of the existent bodies of work on ecocinema), it is also important to contextualize such ideas within the transnational economic and geopolitical systems that structure the world, especially concerning the proposed 'global' solutions to environmental concerns.

To begin, it is necessary to address the oft-used misnomer of transnational cinema as 'world cinema', effectively seen as non-English language films with culturally specific, 'difficult' content, as distinctly other from mainstream European or Hollywood cinema. Much of the normative discourse of Global Hollywood, for example, or of world cinema, perhaps inadvertently, endorses a perspective where 'all endeavours and exchanges are seen either to consolidate Hollywood's hegemony, leading to the centralization of global media and the homogenization of cultures, or to offer resistance, marking the partial triumph, real or imagined, of the local over the global. In either case, culture is being crucially defined in relation to a hegemonic order' (Sarkar 2010: 39). Thus, instead of validating, advertently or inadvertently, the centrality of US-based ecocinema, modifying our understanding of the core set of themes explored in transnational cinema studies allows us to build a more complex picture of the ways ecological considerations work locally.

The eminence of this 'ecological superstructure' nevertheless makes it difficult to avoid reducing representations of other cultures to anything more than a by-product of Hollywood cinema. To navigate this problematic territory, transnational ecocinemas must engage with such concerns in a global framework, providing sustained interrogation of the ways that concerns arising from transnational cinema studies (e.g. hybridity, postcolonialism, diaspora, the geopolitics and economics of a global inequality) feature in the ecocinemas of the world or how cross border narratives mirror the ecosystemic impact of ecological issues. This type of ecocinema engages with issues of uneven global development and sustainability by providing alternative challenges to the formative and narrative conventions dominating mainstream Hollywood cinema.

To take a transnational approach to ecocinema is in its own right a politicized decision, for which we must consider hybridized ideas that challenge the structuring forces of the hegemony—colonial power and human centricism. We must then operate in a dual bind where the cultural-economic centrality of Hollywood cinema is challenged while we also acknowledge that we operate within a global film market that is still largely structured around Hollywood and its global influence. We are also mindful that assuming that Hollywood cinemas operate simplistically as a colonizing force that structure the global cinemap forcibly (though this can also be argued) strengthens this sort dominance. This does not mean, however, that we are not able to propose hybrid ideas that are essential to deconstructing hegemonic paradigms enforcing cultural political relations. A mode of analysis that moves us beyond the existing power relations that structure cinematic production would involve a range of border crossings. This involves not only focusing on discussing critical depictions of global relations but also the introduction of new perspectives that challenge the existing paradigms of ecocinema. It calls for a change to the ways we approach the transnational. Rather than understanding the concept as some simplistic notion of marginal cinema or depiction of cross border activity, we can consider it as a way that allows us to see the cultural embeddedness of all ecocritical rhetoric and their simultaneous need to work beyond such restricting paradigms. This is precisely why transnational approaches to ecocinema are necessary—they enforce the need to rethink the current academic conceptualizations of ecocinema as a form of culturally located production. Instead, they encourage us to see how films both locate ecological concerns in cultural contexts, while they seek to dislocate them from these very same contexts by evoking a more comprehensive understanding of the ecosystem. The processes of location and dislocation are still often done according to the central logic of the nation state framework. By taking a transnational approach to ecocinema, we challenge the centrality of nations in ecological thinking while we also acknowledge that nations persist not only in ecopolitics but also in ecocinema. Through adopting a transnational approach to ecocinema we can start to work towards addressing the inherently anthropocentric and genic logic of film production. Instead of conceptualizing the cultural context of the ecotext as a given, we see it as a constant question.

Transnational Ecocinemas?

Before venturing into more in-depth analysis of a range of case studies that extend the definition of transnational cinema, we must consider the wide ranging implications of Lu and Mi's definition of ecocinema as 'the study of the production and reproduction of life, the relationship between the human body and the ecosystem, and the controlling and administering of the human body in modern capitalist and socialist regimes' (Lu and Mi 2009: 2). If the definition of an ecological film involves texts which reflect the existence of humans in the wider ecosphere, it is very difficult to delineate a non-ecological type of cinema. The field conceivably concerns genres from science fiction to urban crime thrillers, from westerns to fantasy, and, according to Pat Brereton (2005), ecology has been a persistent meta-narrative in popular Hollywood cinema since the 1950s. While a range of Hollywood genres have been discussed by Brereton and other commentators, how does this dilemma look from the perspective of the even more dispersed and varied transnational cinemas? Nick Kaldis' discussion of ecocritical work in Chinese cinema serves to illustrate some of the ways in which these ecocritical texts can be seen to operate:

There is no narrative closure and catharsis, no gesture towards the promise of less disruptive and more locally engaged, ecologically friendly, and sustainable developmental policies. Instead, the traumatizing exigencies of national development are shown to be woven into the very fabric of everyday life, creating irresolvable intra- and interpersonal conflicts, anomie, sexual dysphoria, dislocation, loss etc .... [in] projecting the characters' confusion and disorientation onto the viewing audiences. A work of art can make original and complex contributions toward our understanding of the conflicted state of human agency under conditions of ecological devastation in China today. The unquestionable good of economic 'progress' and 'modernization' is shown to be the cause of irreversible ecological damage and obliteration of historically-rich local communities.

(Kaldis 2009: 72)

While the above description involves China-specific representations of environmental issues, Kaldis' perspective could be applied to a range of films from different national contexts. What makes them relevant for our purposes is that these local or national concerns have clear transnational and global dimensions (especially when it comes to ecological effects). When sociorealist texts such as *Moartea domnului Lăzărescu/The Death of Mr. Lazarescu* (Cristi Puiu 2005) or *Gomorrah* (Matteo Garrone 2008)—which contain a distinct human ecological strand in their explorations of physiological effects urban planning and socioeconomic development have on individuals—are discussed as transnational cinema, this is often conducted in the context of the world cinema connotation of the term.

This is also the case with more explicitly fantastical or anti-realist films such as *Dogville* (Lars Von Trier 2004) and *Mies vailla menneisyyttä/The Man Without a Past* (Aki Kaurismäki 2002), which both make concrete attempts to not only urge us to consider social injustice, but also the ways in which the contemporary social order is creating a sociocultural environment that is not only detrimental to interpersonal relations, but also to human health. Crucially for us, *The Death of Mr. Lazarescu* and *The Man Without a Past* largely work within the constraints of national borders, but they also frequently connect with transnational concerns, if not consistently in sociopolitical representation, at least in artistic terms. Similarly, *Gomorrah* and *Dogville* work transnationally, the first in its construction of what David Bordwell (2008) has called the 'network narrative' (a prominent and increasingly popular strand of transnational cinema focusing on narratives that connect disparate groups of individuals), the second in its Brechtian exploration of American society. These are only some of the types of films that we can explore under the rubric of transnational, where, crucially, environmentalist or ecocritical analysis can yield substantial new ways of exploring these well-known texts. The focus on dialectical tensions on the formative level combined with a pointed engagement with global capitalism is able to create the type of emergent ecocinema, which in Andrew Hageman's view (2009), combines the cinematic, the ecological and the ideological.

My intention is not to make a claim about the superiority of art film as a 'better' form of critical cinema. Instead, the argument is concerned with expanding the scope of ecocritical studies by focusing on films which work from an ecocosmopolitan basis as they construct arguments that refuse the localism of national cinema and the globalist attitudes often seen in the mainstream. To explore these topics, I focus on two films with a transnational scope, which challenge the types of films normatively included as part of ecocinema through their thematic focus, Gaspar Noé's *Enter the Void* (2009) and Jim Jarmusch' *The Limits of Control* (2009). Importantly, both films are transnational in theme and production scope, but they are simultaneously in English and produced by white Western men, albeit ones known for challenging the norms of their societies. Studies of transnational cinemas often criticize the normative centrality of Western culture, concerns which these films reproduce both on the level of production and theme. These films are also appropriate for the current study as they are key examples of what Pam Cook (2010) argues comprises 'the popular art film'. The relatively large budgets of the film, the prestige of their directors, the extensive coverage they receive in international distribution and reception, and finally, their subversive uses of conventions from mainstream Hollywood cinema, all enable them to act as key portals into the study of transnational ecocinema. As we will see, it is these contesting dynamics that allow the films to function as examples of how to mobilize transnational considerations to expand the field of ecocinematic studies. Furthermore, these films are not blatantly environmentalist, yet their thematic scope extends to many of the themes covered in ecocritical literature. This is precisely the point of analysing them—if we only pay attention to explicitly environmentalist texts, we ignore the substantially rich and varied field in which ecocritical arguments can be expressed. In adopting an enthusiastically 'second tier' ecocritical approach, we chart two key ways in which transnational and ecocinematic concerns intersect: the first explores how the urban-human interface works ecophilosophically, the second, the ways capitalism impedes on transnational ecoscapes.

Transnational Urbanism

Urbanization and all its discontents form an essential part of transnational studies and the issue also functions as a key way of expanding the scope of ecocritical cinema. Countless films deal with claustrophobia in the city and the darker side of urbanity. They explore environmental diseases and health conditions, the genetic mutations that occur in the 'wildlife' of cities, the despairing conditions in which individuals live, and the ways governmental and council policies impact on individuals that habit the cities. Concerns of human ecology make themselves explicit in the interpersonal and communal degradation that seems entirely natural in these environments, as almost if these spaces were designed for this sole purpose. Genres such as film noir and body horror metaphorically reflect the socioeconomic realities of cities and the human estrangement these man-made environments create.

In the majority of Hollywood's engagements with nature, it acts as a space for human, often masculine, self-realization. Many ecological films wrestle in order to come to terms with reconciling the inorganic coldness of the city with a more ecological understanding of its position as an extension of humanity's organic and natural position in the ecosystem. One way of overcoming this is to abandon realistic engagement with the cityspace in favour of a more cognitive metaphysical perspective, something which Chris Tong (2009) identifies in Chinese cinema in a film such as *Rainclouds over Wuhan/Wushanyunyu*

(Zhang Ming 1996). Similarly, metaphysics are key to many US-based city films, ranging from *Synecdoche, New York* (Charlie Kaufman, 2008) to *Inception* (Christopher Nolan 2010), the city is morphed to accommodate individual identity and perspectivity, a sort of mechanical-organic interface for self-realization.

Memory, or specifically represented memory, functions as the guiding nodal point in this organized recreation of the city, thrusting and coalescing according to its reproductive stage. The interconnections of the protagonist's consciousness and the diegesis creates a sense of 'urban organicism' (Rozelle 2002: 09), which tells us of the ways in which the human mind responds to the tactile surfaces of the cityscapes and morphs its material surfaces as if they were a form of cybernetic flesh. Crucially in these two examples, individual mastery of the city is emphasized, as the city becomes a soft, controllable element, an extension of man's mastery over the natural environment. A more modified version of this mastery comes through in, for example, the French *District B13* series (2004–2009), where parkour acrobatics enable the protagonists to scale up and across the concrete environment of the city. The urban navigation emphasized by these films is a more pragmatic approach, modifying the rejection implicit in most sociorealist cinema (including genres like film noir) and the anthropocentric approaches of *Synecdoche, New York* and *Inception*. In many ways, the reciprocal elasticity of the protagonists and the concrete world gestures to a more complex and comprehensive understanding of humanity's place in the ecosystem, a conceptualization which sees the ecosystem as a 'breathing landscape that is no longer just a passive backdrop against which human history unfolds, but a potential field of intelligence in which our actions participate' (Abram 1997: 260). But this is still, of course, a man-made environment, now controlled more than ever before by its creator. Working from this metaphysical turn, we focus on Noé's *Enter the Void*, which embodies an emergent ecospace combining both urbanity and the imagined organicity of human space (the habit of anthropocentrically imagining material constructions as malleable organic material) in unexpected ways.

The Spectacle of Subjectivity in *Enter the Void*

The initial attraction of *Enter the Void* is its first-person narrative perspective, a consequence of Noé aspiring to recreate a drug-fueled out-of-body experience he went through while watching *Lady in the Lake* (Milland 1947). We gaze at the world through protagonist Oscar's eyes as we are introduced to the world of the film and his sister Linda and best friend Alex. Oscar's initial marijuana intake propels us into kaleidoscopic membrane-like structures, which dissolve into an out-of-body experience as he observes himself from above. While Oscar's elevation soon comes to an end, the hazy perspective continues as we follow Oscar and Alex from above as they wander through the neon-lit streets of Tokyo. Oscar's drug-laden subjectivity culminates in tragedy as he is shot by the police, suspected of dealing drugs. After his initial befuddlement, the rest of the film follows his spirit as he flashes back in time and space, while we also follow him observe how Linda and Alex coping with his death.

While *Enter the Void* is not a traditional ecological text, its exploration of human subjectivity encountering the corporeality and metaphysicality of existence enables us to explore the challenges of a world increasingly devoid of any sense of organicity. Noé's earlier work *Irréversible* (2000) experimented with immersive techniques as it complemented its visually and thematically shocking content with subliminal sound effects and swooping camera movements intended to have a nauseating effect on the spectator. *Enter the Void* contains many of these ideas as we are no longer bound by the limitations of the body. The persistent and unique maintenance of the subjective perspective enables the film to work on a haptic level as it encourages the spectator to immerse themselves into the world. The immersive out-of-body perspective and drug-induced kaleidoscopes bring to mind the 'Ultimate Trip' of Kubrick's *2001* (1968), and its metaphysical projections of a post-human future. But as with the rebirth of the human race in the form of the star child in Kubrick's film, it seems the mind still craves the body as Oscar is literally lost with his mind flowing randomly through time and space looking for some sort of connection.

The deliberate confusion and ambiguity fostered through city shots, flashbacks, seemingly present-day conversations, and memories, recreated with the back of Oscar's head in the frame, allows us to observe the relationship between humanity and the cityscape in novel ways. The urbanity of the city with its concrete buildings and neon lights creates an unusual atmosphere not unlike the afterlife in *A Matter of Life and Death* (Powell and Pressburger 1947) with its expansive and sterile surroundings. But the sterility of the afterlife contrasts with the life that still perseveres inside the concrete, and Oscar's rebirth and the film's frequent allusions to *The Tibetan Book of the Dead* merge the organic and the inorganic. And slowly, the abstract flow starts to achieve a rhythm through repetition of the aural and visual aspects of the film, as Oscar seeks out his sister who is now sleeping with Alex. The climax of the film evolves into an orgy of thriving bodies and urban surroundings, as the subjective perspective moves through the concrete and the bodies, fusing them into one. Through this, the film emphasizes human corporeality as the only way to navigate through this concrete jungle, while it also indicates a more organic understanding of the city: the city may still shelter humanity, but it is nothing more than an abstract construction if humanity ceases to inhabit it, to give it its validity and reason for existence. The extended copulation segments and Oscar's rebirth underline human interconnectivity as the key to the survival of the human race, questioning the often held assumption of the city as an organic and self-aware entity defining the path of human life. The irony is that it is truly a city-film, but its ecological framework displaces the city into a secondary, if still necessary, part of the contemporary ecosystem.

While *Enter the Void* depicts humanity's attempts to overcome the sterility of the cityscape with more organicist approaches, cities remain the most obvious manifestation of humanity's attempt to reinvent its relation to the natural environment. It acts as a visual material signifier of a mode of thought that elevates modernization and progress as absolute virtues, receding nature into utopianist ideas of 'authenticity'. To take this discussion further, capitalism can be considered the predominant philosophy that organizes this sort of thinking into practice.

Expanding on the ways capitalism and ecological thinking interact in cinema, we explore Jarmusch' *The Limits of Control* and the ways it urges us to rethink capitalism's relationship with the environment. This discussion complements assertions on *Enter the Void*'s urban organicity and sets the stage for exploring the ways both films operate transnationally.

The Limits of Control: Green Politics in a Capitalist Ecosystem

Capitalism is a key focus in a lot of mainstream and experimental forms of ecocinema, ranging from Godfrey Reggio's Qatsi-trilogy (1983–2005) to Steven Soderbergh's *Erin Brockovich* (2000). These films capture what ecocritic Joel Kovel sees as the fundamental obstruction to breaching humanity's self-proclaimed superiority over nature:

> Capital in its essence is not directly part of nature at all. It is a kind of idea in the mind of a natural creature which takes the external form of money and causes that creature to seek more of what capital signifies. It is this seeking through economy and society that degrades nature. Capital becomes both a kind of intoxicating god, and also a forcefield polarizing our relation to nature in such a way that spells disaster.
>
> (Kovel 2007: 4)

It is of course not surprising that capitalism features so centrally as the (ab)uses of resources and the problems of sustainable development are some of the vital concerns explored in ecocinema.

This chapter takes a well-versed topic—corporate corruption—and explores it through the prism of transnational ecocinema. In much of mainstream Hollywood cinema, corporate politics is embodied by a typecast corporate villain—these figures range from Emmerich's films to the CEO of the nuclear power corporation in the remake of the BBC series *Edge of Darkness* (Martin Campbell 2009). The villains (sometimes literally) drip scumminess and moral corruption in their unyielding drive for increased profit. These examples tap into a wider distrust of corporate politics and the frequent scandals involving corporate greed and misdeed (from the financial crisis to Bernie Madoff). The role of corporate politics in environmental destruction is certainly an important part of ecocinema, but painting such a black and white depiction of villainy is problematic as it often acts as a simplification, a deflection, of any sort of wide-ranging and sustained criticism, functioning as a way to put a greenwashing label on a problem that is much more complicated in reality.

The essential argument for environmental economics within the capitalist system is that by privatizing nature, people learn to care for it as their property. However, the problem is that, by being made property, 'nature is thus severed from its ecosystemic existence and becomes subjugated to human agency. Thus, the ceaseless rendering of the environment into commodities, with its monetization and exchange, breaks down the specificity and intricacy of ecosystems' (Kovel 2007: 40). Capitalist maneuvering packages the environment as something that can be exchanged and controlled, and many of the cinematic imaginings, perhaps inadvertently, contribute to this packaging by allowing us to identify easy villains and simple answers to complex questions. By creating pantomime villains with grandiose schemes of world exploitation, or exploring the role of specific organizations and nations in the process of ecological destruction, the problem is made to be one of the 'contemporary system'. And as we are repeatedly told and shown, the system is too big to fail. This logic complements Ingram's suggestion that ambiguous conceptualizations of humanity mask the role sociopolitical power plays in environmental exploitation: 'there can be no general species accountability for ecological damage, due to global inequality and the stratas of exploitation' (Ingram 2000: x).

Jarmusch' *The Limits of Control* provides us with an appropriately subversive perspective on the relationship of capitalism and the environment. The film's elliptical and elusive narrative follows an unnamed assassin (Isaach De Bankolé), who is ordained to carry out an unspecific mission by a group of shady characters in an airport. He only has the following quasi-philosophical instruction to guide him: 'He who thinks he is bigger than the rest must go to the cemetery. There he will see what life really is: a handful of dirt'. As the assassin travels through Spain, we follow him through a series of meetings in coffee houses where he is met by different individuals who provide him with matchboxed instructions. These eventually lead him to a heavily armed compound, where he is able to sneak in 'using his imagination', as one of the characters puts it, and confront the shady corporate figure he was sent to kill. The corporate head, as portrayed by Bill Murray, is constructed as the type of cardboard villain we see in blockbuster cinema, but the way the assassin carries out his task could not be any different from mainstream cinema.

Jarmusch has a reputation for using genre conventions in unexpected ways, and *The Limits of Control* is no different. Adopting the framework of action thrillers, the film plays games with the expected narrative tropes, from the early planning to the actual execution of the assassination. Each of the assassin's meetings involves well-known actors such as Tilda Swinton and John Hurt, who muse on a range of topics from the role of art to individual human greed. These are, of course, key themes of Jarmusch' work and suggest our need to be aware of the self-reflexive connotations of the narrative. While both versions of *Edge of Darkness*, for example, culminate in the demise of the protagonist, Jarmusch' repetitive and intentionally banal use of cinematic genre tropes does away with most trappings that govern spectator expectations of the genre. As the CEO rants about the power of capital and the supremacy of human needs over the environment, it is clear that the corporation is involved in manufacturing environmentally detrimental products. The unnamed organization that hires the assassin effectively takes the place of natural progress as it takes care of those who seek to exploit resources for their own benefits. To illustrate this, the assassin is repeatedly told by his contacts that reality is subjective, and he is effectively granted the power to alter the constraints of reality for his mission. And while this is not much different from the ways natural disasters purify the contemporary political system in Emmerich's films, Jarmusch strips the patterns and rules that govern human life (and dictate the politics of corporations) of their illusionary safety and lays bare the basic mechanisms which control the contemporary social order, thus suggesting the limits of human control over the ecosystem. Through its challenging narrative and the need for the spectator to engage with the ecopolitics of the film beyond the cathartic potential of feel-bad cinema, the film asks us to interrogate capitalism's normalizing potential, unmasking some of the processes (i.e. genre structures and conventions) through which mainstream ecocinema greenwashes ecological awareness.

Turn to the Transnational

This brief discussion of *Enter the Void* and *The Limits of Control* shows how they deconstruct and subvert the linearizing narratives and rhetoric of mainstream ecocinema. But what role do concerns emerging from transnational cinema play here? Starting with *Enter the Void*, its transnational politics emerge from a sense of ontological hybridity in its representation of Tokyo. It combines a range of approaches to the city—real spaces, CGI movements, subjective perspectives, and metaphysical visions—into an immersive experience. It creates a sense of the ecosublime, which according to Lee Rozelle, concerns the following: 'If the sublime, in the Kantian sense, transports literary figures and readers beyond reason, beyond language, into an aesthetic mode of higher finality, then the ecosublime might well embody a wide-ranging reaction to natural environments in rapid decline, of human life struggling to come to terms with the excesses of its materialist progress' (Rozelle 2002: 100). But Noé's insistence on situating the film in Tokyo adds a transnational quality to these ecophilosophical perspectives, which differentiates it from other metaphysically Buddhist attempts at the ecosublime (Aronofsky's *The Fountain*, 2006, comes to mind).

According to Noé, the decision to set the film in Tokyo was to capture the visual stimulation of the city's neon-lit spaces and construct an argument against the restrictive drug laws of Japan. Yet, the Tokyo of the film is far from a topographical reality of the city. The ways the film blurs and merges different aspects from the past with its transient, unbound representation of the present allows the film to rethink the ways the body relates to the inorganic concrete of the city. As Tokyo is transformed into an experiential metaphysical mindscape, it also embodies a typically problematic outsider gaze on what is still a culturally specific space. We never learn anything about the Japanese context beyond a superficial condemnation of its drug culture. Even if the spectator of the film happens to be Japanese, the transformation the city has undergone necessitates viewing it from a crosscultural perspective. The film's transnational politics are thus contradictory: On one hand, it suggests that the cultural-political space of individuals does not matter in a universe based on the logic of reincarnation. Simultaneously, the film is clearly set in a cultural coded space (even if this is an exoticized, 'diluted' one) and depicts the inhumanity invading the Tokyo cityscapes by exploring it from the culturally other perspectives of its protagonists. This contradiction highlights both positive and negative aspects of transnational ecocinema, which need to be able to discuss the relations between global ecosystems and those of capital and culture, while also acknowledging that all cinematic arguments rely on the realities of a world of borders and, often, systemic global inequality. Thus, the westernized outsider gaze on Japan in *Enter the Void* remains highly problematic, but its pan-human concerns created through its de-culturalized city allow us to envision its potential in ecocinematic terms. By placing us in the shoes of the protagonist, we are asked to become more clearly involved in the thematic work of pondering about life and reincarnation, corporeality and transcendent metaphysics. It creates a fictional subjectivity which can make spectators engage with these questions in more dynamic and immersive ways than if they were posed in a more linear narrative. Noé's aims at affectivity and hapticity—the sharing of memories and the emotional connections these can create—thus construct an effective ecophilosophical meditation on corporeality and reincarnation. But inspected from the perspective of transnational studies, we have to remain very critical about these types of depoliticized and abstract forms of ecocinema, despite any progressive steps they may take in philosophical and cinematic terms.

Posthuman Perspectives

How, then, do these abstractions work alongside the transnational qualities of *The Limits of Control?* The reworking of the conventions of cinematic narrative in Jarmusch' film is exactly the opposite of the linear path that Emmerich's *2012* takes. Both films culminate with a demonstration of nature's might—*The Limits of Control* with its protagonist unraveling the physical rules of the universe, *2012* with the US-led coalition withstanding the destructive power of nature in the Chinese-manufactured megaboats. Both films also seemingly engage in leftist political rhetoric with their anti-authoritarian argumentation about the need to rethink social hierarchy to avoid the contemporary state of the world. Whereas *2012's* conclusion validates the status quo—Western political and cultural systems and the military maintain hierarchical domination, including the prescribed leading roles available for Western intellectuals—*The Limits of Control* interrogates the roles language and social customs play in the maintenance of ecological exploitation. The intentionally fragmented and cyclical narrative aims to draw spectators into its deconstructivist impulses and the ways in which conventional revenge narratives are constructed. The frequent question of whether the assassin speaks Spanish (to which he invariably responds no) undermines the hierarchical role language plays in global power politics. This also goes for many of the abstract pseudo-philosophical ruminations that pepper the film, all of which essentially connote humanity's immersivity in the ecosystem. The ruminations act as inverted parodies of the ecopolitical argumentations of mainstream ecocinema, demonstrating the often abstract and insubstantial nature of their greenwashed rhetoric.

Jarmusch has, of course, used such distancing techniques throughout his career, for example in *Ghost Dog* (2000), where characters communicate with one another without understanding each other's language. In much the same way, the absence of shared communication and the abstractions of shared rhetoric highlight their limited significance in contrast to the realities of the ecosystem. *The Limits of Control* highlights the need for instinctual guiding tools in an uneven and unpredictable ecoreality divested of the conventionalities and structuring principles of conventional society. By emphasizing the fragility and fallibility of physical reality, the film suggests that the limits of humanity's control over the environment and other people are not as tangible and powerful as are often conventionally thought. Physical reality can be manipulated for a cause, especially if this cause is to do with undermining the power of capital. By drawing attention to the mechanisms that cover normative ecological discourse, *The Limits of Control* makes a central theme out of the topic of dislocation and the point of cultural veracity in an ecologically unbalanced society. And in its inventive deconstruction of transnational communication and its relationship with global power, it provides an intriguing, if abstract argument to David Pepper's suggestion that it is 'social justice ... or the increasingly global lack of it, that is the most pressing of all environmental problems' (1993, xi–xii).

The Futures of Transnational Ecocinema

For writers such as Paula Willoquet-Maricondi, ecocinema is a term largely concerned with films that seek to motivate personal or political action (2010: 45). It would seem that neither Noé nor Jarmusch' films meet this qualification. Nor do they abide with 'environmentalist cinema' as films which use environmentalist issues as a narrative element to maintain anthropomorphic perspectives on the ecosystem. Rather, they challenge these perspectives all the while using anthropomorphic techniques to conduct their criticism. Meanwhile, their complex ways of involving spectators calls for active engagement with their ecological content, despite the absence of easily identifiable 'environmentalist' content. It is precisely this complexity in their ecological orientation that enables them to operate on a multitude of levels and provide ways to expand the field of cinematic ecocriticism. The politicized call inherent in much of environmentalist cinema is a significant part of the identity of ecocinema as an academic field of study, but it can also be seen as a prime contributor to what Lee Rozelle outlines as a key dilemma facing an ecocritic working within the cultural or media studies tradition:

> Against the postmodern inclination toward unrestrained symbolic and cultural 'play', green literary theory has been depicted as somewhat of a semantic killjoy prone to foster a reconstructive approach to language and a reliance on a 'real' superstructure. It's also common to find green readings that leave the process of representation or psychological states of ecoterrorized minds in positions of primacy.
>
> (Rozelle 2002: 5)

The strive to participate and make change is a problem for ecocritical humanities as they face marginalization by the sciences and scepticism from the general public. Another potential problem concerns the doom-laden or explicitly preachy rhetoric often deployed in key ecofilms which are focused on trying to shock into environmentally aware action, and adaptation of these tactics by academics can lead to a much stereotyped, surface-based conception of the ecocritic—the hippified liberalist enamored with Mother Gaia. Contemporary ecocriticism is certainly widening (or sharpening) its scope into real life political and social considerations without losing the necessary objectivity required of the academic. Critics like Lu and Mi show a more pragmatic approach to ecocriticism instead of merely presenting doom-laden apostrophies about ecocatastrophes (though these also exist in their collection). While political engagement is certainly a necessity for academics working in the field of ecocinema, this should not sideline more philosophical ecocriticism and attempts to engage readers from a range of backgrounds.

Enter the Void and *The Limits of Control* answer some of Rozelle's concerns as they illustrate ways in which transnational ecocinema can offer different avenues for cinematic ecologicalism beyond the panoptic greenwashing of blockbusting ecocinema. To put it bluntly, if Emmerich's ecofilms 'flatten' the world, these films work in the opposite direction to deflatten it. Yet, films are commercial products that aim for the widest possible audiences, even ones with seemingly elusive art house content. Thus, we have to question the cultural role and technological means of both directors as both films are, after all, produced by two Western males well known for their art house credentials and exhibiting a seemingly elusive ecological perspective. If we were to explore the two case studies here as either ecological or transnational cinema, we would end up ignoring crucial aspects of their content and potential. Combining tools from both fields results in an analytical framework that allows us to expand the ways we approach these films, and consequently, the fields in which we work. Taken as forms of transnational ecocinema, the films evoke complex meanings which remain internally contested in their political rhetoric. They retain the primacy of English and the key mediating role of Western artists, while simultaneously, they offer self-reflexive critiques of this perspective by indicating the metaphysical insignificance of sociopolitical organizations and power. Their metaphysical conceptualizations of human life in the contemporary ecosystem provides a more critical and immersive negotiation of ecological themes than available in the feel-bad pleasures of most mainstream ecocinema.

The adoption of a transnational approach for this chapter has meant that I have not focused as much on questions of place as is common in ecocinema studies. But neither have we endorsed the sense of geopolitically centralized planetarism that acts as the cultural political *modus operandi* of Hollywood cinema. The oscillation between the specificities of place and the ambiguities of the planetary is another indication of the ways in which transnational cinema studies can contribute new perspectives to ecoritical writing. By deprioritizing place but still maintaining its organizing presence, transnational ecofilms can overcome some of the limitations of the binary between globalized and national forms of ecocinema. By questioning the rationale of spatial specificity, the films propose a similar rhetorical direction as Heise's ecocosmopolitanism. But this is also a problem as the ambiguity and non-commitment of 'planetary' perspectives can negate the potential of these films to participate in politicized discussions as well as evoke some of its central limitations, such as priviledging Western perspectives. While they are admittedly esoteric—and thus marginalized in the global marketplace, and flawed in their transnational-ecophilosophical argumentations, being too centralizing in their approach to other cultures—they work to unravel the normalizing tendencies of the capitalist mode of production espoused by the Hollywood system. By interrogating the ecological hegemony of the existing cultural industry, these films act as reminders not to take the current status quo of popular ecocinema for granted. Content based discussion of transnational ecocinema may operate on a deterritorialized planetary level, but the type of 'critical transnationalism' espoused by Will Higbee and Song-Hwee Lim (2010) act as constant reminders to interrogate the costs and politicized implications concepts such as the planetary and the ecocosmopolitan bring to these analyses.

Moving beyond Ideology: Future Directions for Transnational Ecocinema

While textual discussion of transnational ecofilms is a crucial starting point for challenging the dominant paradigms of ecocinema, future studies need to consider their production and circulation in depth. While few reviewers have discussed *Enter the Void* or *The Limits of Control* in ecological terms, transnational approaches to reception studies can yield unexpected results for films that seem to operate primarily within the national paradigm. Jiayan Mi suggests that films such as *Sanxia haoren/Still Life* (Jia Zhang-Ke 2007) challenge the spectator by refusing visual spectacle and choosing instead to focus on destruction and depletion of the environment (Mi 2009). Taken transnationally, the ecological work of the film can be considered from an alternative perspective, where the disappearing spaces of the Three Gorges work as a form of exoticism, of seeing the primal ways in which the 'other' slowly modernizes, and the costs this takes. The spectacle of emotional destruction (Jia's film is entirely devoid of the manufactured CGI-spectacle found in Emmerich's films) arguably evokes a feel-bad reaction in the Western spectator, which mirrors the ecocatharctic effects seen in Hollywood cinema. While the 'naturalization' of non-Western cultures as primal others is a handy way for satisfying the critical curiosity of the Western spectator, *Still Life* is clearly not to be dismissed as an imitation of Hollywood ecocinema. And indeed, why should we only consider Western audience perspectives? Arguably, it would be as, or even more, important to consider the ways audiences in all parts of the globe utilize both mainstream blockbusters and more local fare. Regardless, *Still Life* is not popular cinema in its domestic market and remains marginalized compared to Emmerich's films. In a study conducted by the author (Kääpä 2012) on the visibility rate of *Still Life* in comparison to *The Day After Tomorrow* with Chinese university students, *The Day After Tomorrow* was cited nine times out of ten in response to the question: 'can you provide an example of environmental cinema?' In comparison, *Still Life* was only cited by approximately two students out of ten, and many of the sampled participants had never heard of the film. There is a range of reasons for the marginal status of Jia's film, including the central government's abjection to any film project that contains explicit criticism of its developmental policies. But even after screening the film to students, many of them still found problems with identifying the specific ecological content of the film as, for them, *Still Life*'s politicized participation is more about class politics rather than environmental argumentation. This is in direct contrast with Western reviewers, for example, who without fail equate the film with ecopolitics.

While we do not have sufficient space to engage with Chinese or other audiences in depth here, this brief discussion illustrates the disparate playing field in which different ecological perspectives 'compete', while consumers in different cultural contexts may be increasingly aware of environmental issues. This does not imply that they seek out products that are more specifically about their immediate contexts, or engage with them in similar ways to Hollywood ecocinema. Spectator experience is a key part of the analytical reorientation of transnational ecocinema for two main reasons. First, much of contemporary transnational cinema studies urge consideration of audience perspectives in global cultural politics (see Durovicová and Newman, 2010, for examples of translocal reception studies). Second, understanding the cognitive and political processes involved in ecospectatorship will allow producers and analysts to develop more incisive and impactful methods for fostering ecological awareness, a notion which extends also to academic work on cognitive cinema. Engaging transnational audiences can thus challenge the normative 'ecological potency' an academic finds in a given text and contribute to the diverse ways in which these films create their ecological meanings.

Exploring the material footprint of transnational film production provides additional ways in which critical engagement with transnational ecocinemas can benefit both transnational film studies and ecocriticism. While Emmerich's use of green production methods can be considered as greenwashing to merely give the illusion of safer and sustainable production, the transnational production methodologies of more esoteric fare by Noé and Jarmusch are not immune from being ecologically problematic. In fact, transnationalism is by its nature compromised in its ecologicalism. The material costs of travel and transportation will have significant implications for the footprint of even relatively small scale productions as these. The digital recreation of Tokyo in *Enter the Void* can be considered as a way to minimize production costs and highlight the importance of synergizing ecocriticism and transnationalism with convergent modes of media communication. While digital media brings developments in conceptualizing sustainable and ecocritical cultural production, distribution, exhibition and consumption, we must also be critically aware of the ecological footprint of digital media. Digital distribution of e-films was long considered 'environmentalist', but as shown by Sean Cubitt (2009), it has its own substantial detrimental effects on the environment. While developments in green media technology will continue, they may simultaneously have the adverse effect as constant 'progress' necessitates that outdated production and distribution devices are discarded. In addition to this, server farms consume energy to such an extent as to cause pollution in a way that has overtaken even some more notorious producers of ecologically harmful substances. Thus, merely changing the form of the material used will have little meaningful impact. While we must take into account the complex and unpredictable modes in which cinemas circulates nowadays, we must also remember that any attempt to capture the current state of ecofriendly communications technology in academic forms of writing faces an uphill battle from constant, unexpected developments and the industry-mandated efforts to maintain the accelerating returns from the capital invested in these productions.

Finally, we must not assume that media will be consumed in the orthodox ways the cultural industries have traditionally intended them to be used, or expect them to be mobilized in only those functions which may satisfy conventional understanding of the materialist-environmental potential of media. As a very brief example, farming communities on the Lamma Island in Hong Kong frequently use DVDs as scarecrow devices to protect their crops. On one particular visit, we spotted both Hollywood blockbusters, *Around the World in 80 Days* (Frank Coraci 2004), and popular Hong Kong cinema, *2 Young* (Derek Yee 2008), used in this unorthodoxically ecological way (see Figures 1 and 2).

Beyond their commercial runs in theatres and DVD distribution, they first discover second life as pirated discs containing downloaded material, and then as a means to cultivate crops in sustainable and often chemical-free small-scale farming (though the use

Figure 1: Farming communities at Lamma Island.

Figure 2: DVDs put to alternative use.

of indium as a surface coating of the discs has substantial detrimental effects on recycling) (see Cubitt 2009 for more on these complications). While this ecological pragmatism is thus limited, it does underline the extent to which ecocinema can be much more than feel-bad greenwashed blockbuster fare or metaphysical/philosophical advances in our ecological thinking. While the transnational circulation and critical discussion of these films and the increasing popular awareness of their messages make important contributions to advancing global sustainability, the second life these films may find in the form of farming tools reminds us of the materiality of all cinema production, even in the era of the so-called digital revolution. Whatever the case, synergizing concerns from transnational cinema and ecocritical studies enriches both fields and provides new avenues of interrogation into the societal potential of cinema. As we can see from the different locations we have covered in this chapter, transnational concerns hold both a place centric dimension while they allow us to consider philosophical/societal concerns beyond their immediate geographical confines. They necessitate transnational scholars to consider the footprint the films leave while they encourage ecocritics to engage with societal and cultural politics in ways that challenge the occasional abstractions of ecocriticism. Through this, both fields will need to considerably diversify their methods of interrogating the fundamental tenet of each field—their social and political commitment.

References

Abram, P. (1997), *Spell of the Sensuous*, London: Vintage Books.

Brereton, P. (2005), *Hollywood Utopia: Ecology in Contemporary American Cinema*, Bristol: Intellect.

Buell, L. (2005), *The Future of Environmental Criticism: Environmental Crisis and Literary Imagination*, Oxford: Blackwell.

Cook, P. (2010), 'Transnational utopias: Baz Luhrmann and Australian Cinema', *Transnational Cinemas*, 1: 1, pp. 22–35.

Cubitt, S. (2005), *EcoMedia*, Amsterdam: Rodopi.

—— (2009), *Ubiquitous Media, Rare Earths: the Environmental Footprint of Digital Media, Pervasive Media Lab*, Bristol: University of West of England. http://homepage.mac.com/ waikatoscreen/talks/UWEnotes.pdf. Accessed July 20, 2011.

Durovicová, N. and Newman, K. (eds) (2010), *World Cinemas, Transnational Perspectives*, London: Routledge.

Foucault, M. (1995), *Discipline & Punish: The Birth of the Prison*, New York: Vintage Books.

Friedman, T. (2007), *The World Is Flat: A Brief History of the Twenty-First Century*, New York: FSG.

Gardels, N. (2008), 'Post-globalization', *New Perspectives Quarterly*, 25: 2, pp. 2–5.

Hageman, A. (2009), 'Floating consciousness: the cinematic confluence of ecological aesthetics in Suzhou River', in S. H. Lu and J. Mi (eds), *Chinese Ecocinema: In the Age of Environmental Challenge*, Seattle: University of Washington Press, pp. 73–92.

Heise, U. (2008), *Sense of Place and Sense of Planet: The Environmental Imagination of the Global*, Oxford: Oxford University Press.

Higbee, W and Lim, S. (2010), 'Concepts of transnational cinema: Towards a critical transnationalism in film studies, *Transnational Cinemas*, 1: 1, pp. 7–21.

Hwang, S. (1998), 'Ecological panopticism: the problematization of the ecological crisis' *College Literature*, 26: 1, pp. 137–49.

Ingram, D. (2000), *Green Screen: Environmentalism and Hollywood Cinema*, Exeter: University of Exeter Press.

Kaldis, N. (2009) 'Submerged ecology and depth psychology in Wushan Yunyu: aesthetic hindsight into national development', in S. H. Lu and J. Mi (eds), *Chinese Ecocinema: In the Age of Environmental Challenge*, Seattle: University of Washington Press.

Kovel, J. (2007), *The Enemy of Nature: The End of Capitalism or the End of the World?* London: Zed Books.

Kääpä, P. (2012), 'The politics of viewing ecocinema in China', *Interactions: Studies in Culture and Communications*, 2: 2, pp. 145–62.

Lu S. H. and Mi J. (eds). (2009) *Chinese Ecocinema: In the Age of Environmental Challenge*, Seattle: University of Washington Press.

—— (2009), 'Framing ambient unheimlich: ecogeddon, ecological unconscious, and water pathology in new Chinese cinema', in S. H. Lu and J. Mi (eds), *Chinese Ecocinema: In the Age of Environmental Challenge*, Seattle: University of Washington Press, pp. 17–38.

Murray, R. and Heumann, J. (2009), *Ecology and Popular Film: Cinema on the Edge*, New York: SUNY Press.

Pepper, P. (1993), *From Deep Ecology to Social Justice*, London: Routledge, pp. xi–xii.

Rozelle, L. (2002), 'Ecocritical city: modernist reactions to urban environments in Miss Lonelyhearts and Paterson', *Twentieth-Century Literature*, 48: 1, pp. 100–115

Sarkar, B. (2010), 'Tracking "global media" in the outposts of globalization', in N. Durovicová and K. Newman (eds), *World Cinemas, Transnational Perspectives*, London: Routledge, p. 39.

Tong, C. (2009), 'Toward a Hong Kong ecocinema: the dis-appearance of "nature" in three films by Fruit Chan', in S. H. Lu, and J. Mi (eds). *Chinese Ecocinema: In the Age of Environmental Challenge*, Seattle: University of Washington Press, pp. 171–94.

Willoquet-Maricondi, P. (2010), 'Shifting paradigms: from environmentalist films to ecocinema', P. Willoquet-Maricondi (ed.), *Framing the World: Explorations in Ecocriticism and Film*, Charlottesville: University of Virginia Press.

PART II

Documentary Politics and the Ecological Imagination

Colourful Screens: Water Imaginaries in Documentaries from China and Taiwan

Enoch Yee-Lok Tam

Introduction

The remarkable growth of both China's population and economy over the past several decades has come at a tremendous cost to the country's environment. Desertification in the north-west, pollution in rural areas from chemical plants that have been moved farther away from cities, and various forms of pollution in major urban areas such as Beijing and Shanghai are only small parts of the picture. Indeed, the major environmental threat in China continues to be the supply of water. In 'China and water', Peter H. Gleick notes that China's available water supply per capita from 2003 to 2007 was 2138m³ per year, which is approximately one-fifth of the US average water supply while '[i]n 2006, nearly half of China's major cities did not meet drinking-water quality standards, and a third of surface-water samples taken were considered severely polluted' (2009: 83). The World Bank, on the other hand, projects that in 2030, China's annual per capita water supply will decrease to 1750m³ (2006: 5). These two figures are sufficient to illustrate the dire situation of mainland China's water supply.

On the other side of the Formosa Strait, figures show that Taiwan's water supply and quality is considerably better than that of China. In 2010–2011, 3156m³ per capita of water was available for Taiwan's population while more than 70 per cent of river water in Taiwan was not considered polluted' (Pinsent Masons 2011: 182–86). However, Taiwan is vulnerable to water-related environmental disasters. In 2009, typhoon Morakot wrought catastrophic damage throughout Taiwan, leaving 461 people dead and 192 missing, and roughly NT $110 billion (USD $3.3 billion) in damages. Apart from natural disasters, man-made disasters also cause serious damage to the environment. In 2001 the *Amorgos*, laden with 60,000 tonnes of iron ore, suffered engine failure and grounded on rocks off the southern tip of Taiwan. An estimated 1000 tonnes of fuel oil spilt into the sea. The southern coastal area of Taiwan was seriously damaged by the spill. Yet, oil spills are not the only source of man-made devastation. For example, the coral reefs have been ravaged by tourism, which increases the amount of damage that Taiwan sustains during tsunamis.

Based on this situation, scholars from mainland China and Taiwan are currently working on alternative ways of managing water resources and risks of natural and man-made disasters, offering policy suggestions in order to increase water supply, and improving the efficiency of water usage, as well as enhancing environmental protection and enforcement. However, environmental crises always extend beyond policy issues and resource management. Culture is one of the key issues where these concerns can be addressed and documentary film is one of the nodal points where cultural matters can be effectively engaged. Bill Nichols points out that

> What we speak about in documentary then are those subjects that engage us most passionately, and divisively, in life. These subjects follow the pathways of our desire as we come to terms with what it means to take on an identity, to have intimate, private connection to an other, and to belong in the public company of others. Personal identity, sexual intimacy, and social belonging are another way of defining the subjects of documentary film.
>
> (2001: 75)

While Nichols concentrates on investigating the aspects of personal identity, sexual intimacy and social belonging in documentary film, this chapter intends to extend the discussion to the representation of nature and, for the purposes of this chapter, water.

Recent representations of water in Chinese films highlight the absence or disappearance of water and the dialectics of national modernization and development, as Jiayan Mi (2009) and Sheldon H. Lu (2009) have argued. However, this chapter finds that the representations of water in documentaries from China and Taiwan portray a picture that is quite different from the theoretical arguments of Mi and Lu. Water is represented by different colours, which can be codified, exemplifying a system of symbolic values, just as Mi does as he paraphrases John Gaga's *Color and Meaning: Art, Science, and Symbolism* (1999), mapping colours with their specific connotations: 'yellow' for the rulers (the yellow emperor), 'red' for revolution and 'black' for anti-revolutionary acts in Chinese culture (Mi 2005: 334). Following from Mi's work, this chapter maps the colours that appear in documentary films from China and Taiwan and contextualizes their specific connotations in the context of screen ecocriticism and water imaginaries. In this chapter, screen is not only a medium for projecting film images, it is an instrument to show the different aesthetics brought by different colours while it also operates as a device for interrogating the ecological relationship between humanity and water (or nature). With a particular colour, a specific ecological relation can be yielded. By giving a brief historical account of documentaries from mainland China and Taiwan concerning water networks, rivers (Huang He and Changjiang)[1] and the sea (Taiwan as an island), this chapter, starting from the 'yellow screen' of Huang He, illustrates how the 'green screen' relates to socialist ecology to form what the chapter refers to as a green screen with Chinese characteristics.

From here, we explore the 'colourless screen', which isolates water from rivers to help imagine the nation in terms of a national water network and as a modern engineer. The 'blue screen' of the undersea world, in turn, engenders a specific form of ecosophy, which combines the material,the technological and the social. Finally, we come to the 'black screen' of the transnational sea imaginary, which proposes and oscillates between internationality and ecological postnationality.

Green Screen with Chinese Characteristics

To understand the evolution of water imaginaries, we start from a set of mainland Chinese TV documentaries produced by China Central Television (CCTV) from the 1980s to the 2000s.[2] While China is internationally renowned for its independent documentary film movement, TV documentaries dominated the screen in the 1980s, which correlated with the *gaige kaifang*/Reform and Opening Up policy proposed by Deng Xiaoping. This policy has greatly altered how nature and the surrounding environment being represented. At the time, watching TV was the only form of media entertainment available to people. Although CCTV has nowadays lost some of its influential power to the development of new media and local TV broadcasts, in the 1980s CCTV was the dominant channel through which the state conveyed its ideological propaganda. Despite the apparent aesthetic inadequacies, CCTV productions provided audiences with impactful representations of the Huang He and Changjiang as well as a national water network system, which documentary film-makers outside of the official system were unable to achieve.

While the polemical six-episode television documentary series *Heshang/River Elegy* (CCTV 1988a) received much publicity in its time, a contemporaneous documentary series *Huanghe/Yellow River* (CCTV 1988b) is of more significance to the discussion of water imaginaries. Unlike *River Elegy*, *Yellow River* was a large-scale production that consisted of 30 episodes. It began with the discovery of the river's origin and ended with its outlet. It followed the usual narrative structure of large-scale productions such as *Huashuo Changjiang/Discovering Changjiang* (CCTV 1983) and *Huashuo yunhe/Discovering the Beijing-Hangzhou Grand Canal* (CCTV 1986). The omniscient narrators, usually one male and one female, elucidated the scene, conveying a specific interpretation of the images.

Unsurprisingly, *Yellow River* projected the river as the cradle of Chinese civilization. The river was shown to be the source of life as well as of the ethnic specificities of the Chinese people, and thus as the origin of Chinese history. One may discover similarities between this large-scale TV production and *xungen* literature/movement as both of them search for cultural roots via nature. *Yellow River* and the *xungen* movement both emphasize 'the worship of nature as a new religion or ethics [that] may inform the imaginary filling of the nation's spiritual vacuum as it arises anew from the ashes of ideological and religious atheism at the bankruptcy of Maoism-Marxism' (Wang 1996: 219). While the logic of the *xungen* movement is to substitute the national spirit with the worship of nature, *Yellow River* reverses this logic by seeing the river as a symbol of the Chinese race and the nation. The ideological geo-imaginary proposed by *Yellow River* is that water flows from different tributaries and finally merges into a single stream and serves as an ideological support for the sovereignty of China over other ethnic minorities, much like the tributaries that finally merge into the sovereignty of the People's Republic of China.

In the first few episodes, the colour yellow dominates the screen: although the river water is limpid as it flows down from the Tibetan Plateau, it becomes yellow passing through the Loess Plateau, carrying away its soil. Yellow, in mainland Chinese cinema, usually designates the barrenness and backwardness of the country, as in *Huang tudi/Yellow Earth* (Chen Kaige 1984). But in *Yellow River*, the yellow is understood in a natural perspective, if not an ecological one, and resonates with Mi's notion of a 'yellow screen': a yellow screen signals an ecologically deteriorating earth caused by an anthropocentric abuse of the ecosystem and the unsustainable use of natural resources (Mi 2009: 37). Episode 5 'Chuanxing xiagu'/'Via Canyon' focuses on how water flows down from the Loess Plateau, via a canyon to a plain. Intriguingly, the yellow screen becomes green: the narrators recount that this water has been used for centuries to irrigate farmland and turn China into a rich agricultural country. To underline this point, green and idyllic landscapes wash across the screen. By juxtaposing the yellow and the green, the episode concludes that what made the river yellow was the loss of the green: the yellow soil was constantly eroded without protection from trees on the riverbanks. While Mi's version of the 'green screen' celebrated a 'watershed consciousness' that 'advocates a place-based ecological understanding of the interconnectedness between the biotic community and its abiotic environment', and 'toxic consciousness', which alerts people to 'the limits of natural resources and the dangers of both chemical and man-made pollutions' (2009: 37), the green shown in *Yellow River* mourns the fact that the Chinese people had not fully utilized the *Huanghe*'s resources. At the end of the 11th episode, 'Tiansha jintou'/'The End of Sand', the narrators grieve for 'the revenge of nature' by boldly stating:

Green is the colour of human civilization;
Green is the colour of the future;
Huang He is now stridently calling,
calling for the green, calling for the most beautiful colour of the planet.

The poetic ending of the episode seems to advocate an ecological future, but what the 'green' designates here is ambivalent. On the one hand, the narrators point out that human beings and nature are opposed to each other and that in this situation nature will take revenge on human beings, as was the case with the desertification that was happening at the time. On the other hand, the narrators answer the question 'who understands Huang He' by pointing out that Huang He is full of resources and energy. But crucially, the energy here does not refer to the vitality of nature but the potential for water to generate electricity. The ending implies that whoever can fully utilize the energy and resources of the Huang He will truly understand the river.

Rather than Mi's green screen, this version of green is closer to the socialist ecology that appears in Wang Ban's analysis of Su Li's *Women cunli de nianqing ren/Young People in Our Village* (Su Li 1959). Wang differentiates a modification of nature for human needs from a war against nature. He argues,

> The film addresses [the] question by depicting how the organic relationship between humans and nature plays out, and how the collective implementation of the project proves vital to the human 'improvement' of nature—not at the cost of depleting nature but for achieving a balanced form of resonance between human life and the environment. Since the project is conducted by the village as a collective, based on collective wisdom, it aims at altering nature for the common good and for the village's continued survival.
>
> (Wang 2009: 160)

Hence, the socialist model of human–nature relations 'is premised on the collective ownership of natural assets and on farmers as co-producers of a human habitat in tune with the natural environment' (Wang 2009: 161). The difference between Mi's green ecology and Wang's socialist ecology is the position of the human actant. In Mi's ecology, the human actant works within the limits of nature (or natural resources), while in Wang's interpretation, the human actant works upon the limit. While Mi's green ecology does not state clearly the nature of the human actant, Wang's socialist ecology is not ambivalent about this—the actant is the village, a collective of human beings. This rhetoric is taken one step further by *Yellow River* where the actant is the nation itself. Thus, the human–nature relations of *Yellow River*'s socialist ecology expand to an enforcement of nation–nature relations.

The final episodes of *Yellow River* and *River Elegy* further emphasize the ideological implications of these differences. They both show images of the same delta but conclude with different final images of the river. *River Elegy* uses a high, wide shot to show the combined area of the river and sea. The screen is divided by water into two colours: yellow river and azure sea. The yellow and azure represent the land-based, conservative and backward-looking China and the aggressive, scientific, democratized and modern West, respectively. Throughout the entire series, the symbolic connotations of Huang He are undermined, but not without ambivalence. It is, at first, denounced as traditional, superstitious, conservative and sedentary, but near the end of the series, it is praised as possessing the vitality to break through the sedentary characteristics of the Chinese people, leading them into the Azure Sea. By undermining Huang He with a particular ambivalence, *River Elegy* projects a twofold river construction: the river is a demon as well as an agent; it has become a sign of China's backwardness in the past, but can now be a channel that allows the Chinese people to merge with the sea; i.e. the 'occidental' world of science, democracy and capitalism.

In contrast to *River Elegy*, *Yellow River* shoots the merging of the two waters in a much different manner. It does not present the combined area with a high, wide shot. Instead, the camera is placed just above the water level and the shot washes in and out of the waves. There is no colour spectrum; no surge or rigorous flow of water, only a gentle tide moving back and forth. From its origin in the Tibetan Plateau to the Bohai Sea, the river acts as a vein linking the inland and the sea, and as an actant, other than the human actants. In the final episode 'Dahe dongqu'/'East, the River Flows', the narrators account for the transformation of the terrain, as each year tonnes of sand drain into the sea and the land *yanshen*/'stretches out' to it. The description is anthropomorphic since it personifies the sand as if the sand has the will to reclaim the land from the sea. Simultaneously, it shows ignorance of the draining of sand as an ecological phenomenon or a consequence of deforestation. Nevertheless, it affirms that the natural entity, the river, plays a crucial role in shaping the outline of the seashore by turning it into a fertile land. Assuming that the sea carries the same connotation in *Yellow River* as in *River Elegy*, the ending shots of *Yellow River* indicate that China has never been isolated from the world. Instead, year by year, it drains tonnes of sand into and reclaims land from the sea—gradually approaching the world with its own version of a socialist ecology.

In short, *Yellow River* represents water as an element of the river's flow while the river acts as a vein linking the inland to the sea, and its geo-imaginary serves as an ideological support for the sovereignty of China. Based on this, it provides the audience with a socialist model of the nation–nature relationship in which humans utilize nature's resources for national needs. However, this is not simply a reiteration of the older version of the nation–nature relationship from Mao's era. On the contrary, the nation–nature relationship is renewed in the Deng Xiaoping era—working upon expanding the borders of nature as well as approaching the world with this specific version of a socialist ecology. The series provides the audience with a socialist green screen, or, what I refer to as a green screen with Chinese characteristics.

Colourless Screen: Isolation of Water and National Water Network

While *Yellow River* was received poorly at the time of its release and is now largely forgotten except for one- to two-page descriptions in volumes on the history of television documentaries in China (Shi 2000: 76–77), *River Elegy* was well received and triggered a rigorous discussion on Chinese civilization. Yet, the aquatic geo-imaginary as a national focus ceased to act as a source of material for the official television system after the Tiananmen Square protest of 1989 (Connery 2001: 196). According to Zhang Xudong's periodization, the post-Tiananmen decade is the decade of post-socialism that spans not 10 but 12 years, starting from the student protest in Tiananmen Square on 4 June 1989 to Deng Xiaoping's southern tour in 1992, traversing the Taiwan missile crisis in 1996 and the handovers of Hong Kong and Macao in 1997 and 1999, respectively, and ending with the controversial entry into the World Trade Organization in 2001 (Zhang 2008: 1). An emblematic event is missing from Zhang's list of this 'decade', one which is concerned with water and water documentaries—the inauguration of the Three Gorges Dam project in 1994. Consequently, it reinitiated large-scale water documentary productions concerning *Changjiang* and the Three Gorges project such as *Sanxia beiwanglu*/*Memorandum of the*

Three Gorges Project (CCTV 1997) and the 33-episode *Zaishuo Changjiang/Rediscovering Changjiang* (CCTV 2006), both released on CCTV. In addition, local television stations also produced large-scale river documentaries that told the stories of regional rivers such as the Songhua, Huai and Amur. These documentaries posed new questions ranging from cultural politics—how to *qiangjiu*/'rescue' cultural legacy during the construction of the Dam—to policy on relocation and evacuation in different phases of the construction. However, these new productions do not show any significantly new water imaginaries. Rivers, regardless of locations and names, are always presented as the cradle and 'mother river' of national or regional civilization.

In contrast, the widely known New Documentary Movement produced many independent documentary films outside the official system around the same time. Some of the documentary film-makers were also concerned with the Dam project and produced films addressing these issues directly, such as *Nujiang zhisheng/The Voice of Nujiang* (Shi Lihong 2004), *Chenmo de Nujiang/The Silence of Nujiang* (Hu Jie 2006), *Bingai* (Feng Yan 2007), *Yanmo/Before the Flood* (Li Yifan and Yan Yu 2004) and *Yanmo 2: Gongtan/Before the Flood II: Gongtan* (Yan Yu 2008). Lu Xinyu argues that participants in the movement 'felt the need to go to the grassroots, and to understand China's changes and their causes from [this] vantage point' (2010: 19), rather than from on high. Thus, these documentaries are more focused on social issues than their 'official' counterparts. For example, Feng Yan's *Bingai* focuses on resettlement issues and the struggle between local peoples and the officials, with the rivers inhabited by the protagonists act only as background context for the social issues (Yan 2010: 27–47). Thus, independent documentary film-makers do not provide the audience with a new representation or concept of water. The colour of water alternates between yellow and green, while the symbolic role of the rivers remain to exist either as the nurturing mother or a repressed object suppressed by China's modernization.

Yet, the continuous occurrences of flooding and shortage of water in different regions of China allow documentary film-makers to think beyond the rivers. In addition to the alleged solutions provided by the Dam project, water availability and quality continue to become more visible issues in contemporary China. Because of this, water per se, rather than as an element of the river, enters the screen. The eight-episode *Shuiwen/Questions Concerning Water* (CCTV 2008), hereafter *Questions*, is an excellent example. Each episode attempts to answer a question concerning water in China. The questions are:

Episode 1: When will the water crisis come?
Episode 2: Is the drinking water clean?
Episode 3: Do species have peace?
Episode 4: How can rivers be calmed?
Episode 5: Whose water is it?
Episode 6: How large can the national water network be?
Episode 7: Why is it so difficult to conserve water?
Episode 8: Where will the answers come from?

Based on the questions posed by the series, one can easily identify its primary theme: that of the steady decrease of the water supply and the deterioration of water quality. The representation of water in the series differs from the previously discussed river imaginaries in several respects as it creates a colourless screen, which stands in contrast to the green screen of the earlier productions. First of all, water is not depicted as a constituent part of a river but as water per se. In *Questions*, water seems to have its own identity, which is lacking in the previous productions. Water drops independent from individual rivers frequently appear on the screen and remain unassociated with particular colours. In *Discovering Changjiang* or *Yellow River*, for example, episodes begin with several overhead wide shots of the (yellow) rivers. But in *Questions*, episodes begin with close-ups of colourless water drops falling onto a clean water surface with ripples emanating out from the drops.

Accompanying this colourless representation of water are new concepts about water: the independence of water makes discussions about the footprint of water possible and enables viewers to perceive water virtually. This brings into focus the second difference. Water in *Questions* is not only conceived as belonging to (the feminine) nature, as part of the cultural dichotomy between nature and civilization, but is regarded as a nodal element that permeates both the natural and cultural environments. This idea figures prominently in the third episode's presentation of water pollution in describing how human beings, animals and plants are interconnected in a single ecosystem, with water pollution as a key factor that is altering the local ecosystem and threatening the lives of villagers. The green in the episode is no longer promising or as part of the socialist ecology. Ironically, the colour, usually associated with pastoral, idyllic and harmonic life in rural areas, now represents the devastation of the local ecosystem.

To emphasize this devastation, the episode displays the thriving of water hyacinths each summer in rivers near Shanghai and Hangzhou. For the water hyacinths to thrive, certain nutrients are crucial. These nutrients are supplied by pollutants from animal husbandry, chemical fertilizers and pesticides. Hydrophytes in the rivers can digest certain amounts of the pollutants, but they cannot withstand the over pollution of the rivers. The remaining pollutants/nutrients help the water hyacinths to grow to green the river surface with their leaves. The scene at first appears to be a pastoral landscape painting, but soon the rapid growth of this single species destroys the biodiversity of the ecosystem. When the green prospers, other living things die. Thus, green here stands for threats to the ecosystems, while the lack of colour becomes the ideal colour for good water: colourless indicates no pollution and a balanced and healthy ecosystem. If the green of *Yellow River* suggests a complete utilization of natural resources, pushing the limits tolerated by nature, then the colourless water in *Questions* suggests an understanding of the formation of local ecosystems through their interconnectedness with humans, agriculture, industry, animals and plants.

The idea of water permeating both natural and cultural environments is further developed into a new geo-imaginary as later episodes of *Questions* build the *Nashui beidiao gongcheng/* South–North Water Transfer Project into a networking geo-imaginary. The water network proposed by the series suggests that the entire nation is connected by rivers and canals, a national understanding of an abstract and decentred water network that provides people with clean and sufficient water. This river network is portrayed as being comparable to electricity, transportation and telecommunications networks. Unlike the geo-imaginary of earlier documentaries using rivers as ideological support for the sovereignty of China, this geo-imaginary has no central merging point, and comes to suggest a rhizomatic network, which is open and connectable in all of its dimensions. It acknowledges the fact that people do not only live along the riverbanks but scatter all over the nation. Thus, it is not enough to address centralized water issues, but address regional concerns over the state of water in China. Through this, *Questions* not only acquires a more advanced ecological knowledge, which understands the interconnectedness and interdependency of humans and nature, but it also projects a more comprehensive picture of the nation, which sees provinces as nodal points of a network while the central government is the mastermind engineer expanding common local issues to the national level.

In short, *Questions* proposes a colourless screen, on which water is perceived as water *per se* rather than as an element of a river. The isolation of water from rivers makes water quantifiable and makes the concept of virtual water and a water-footprint possible. *Questions* also proposes a decentred, networked geo-imaginary, which displays knowledge of the mechanism of the local ecosystems and a rhizomatic imagination of the national water network. However, this imagined water network may not be able to emancipate humans and nature. On the contrary, it proposes to impose a nation-made network upon the natural landscape and alter the landscape according to the national will. Unlike the green screen in which humans are actants who work upon the limits of nature, the colourless screen suggests that the nation should be the modern engineer, moulding nature according to its mega plans and needs.

Blue Screen of the Undersea World and Its Ecosophy

The representation of water in mainland Chinese CCTV productions always stops at the borders and deltas, and consequently solidifies the notion of nation. Rivers and water drops dominate the screen while sea imaginaries remain peripheral. The remarkable sea in *River Elegy* only serves as a version of 'Occidentalism', which means 'an oriental imaginary construction of the Occidental Other' (Mi 2005: 332; see also Chen 2002). However, Taiwan documentary film-makers offer a different form of imagined sea by documenting the undersea world with its deep blue colour. A 'blue screen' not only shows another colour of the world, but also brings about a new aesthetics and mechanics of existence (see Past 2009: 57). In this section, Taiwanese documentary film-maker Ke Chin-yuan's work will demonstrate how people move within and interact differently with the environment of the undersea world. Through this, a new, revised assemblage of aesthetics and mechanics of existence can be imagined.

A pessimistic undertone permeates Ke Chin-yuan's work, such as *Amasi/Amorgos* (2003), a documentary about the environmental aftermath of the Amorgos oil spill in 2001. In *Chanfang/Squid Daddy's Delivery Room* (2007), he continues to explore the cooperative sea imaginary, which can be found in the work of researchers and activists, as he works on forging his unique style of blue screen by recording ways to provide a better environment for sea creatures. Ke's documentaries provide a very specific function to 'activists', by whom he means a wide range of people working at the intersections of communication between the modern and the native. In *Jiyi Shanhu/Remembering Coral* (Ke Chin-yuan 2004) he portrays a coral world and the differing attitudes towards the sea held by modern people, native people and activists. To him, modern people, exemplified by developers and tourists, are mobilized by the logic of capital: developers appear wherever a profit can be turned; tourists go wherever leisure can be achieved. Their relationship with the sea is exploitative and devastative, and seldom brings anything positive to the environment. On the contrary, the Taiwanese aboriginals, the *Tao*, are represented as nature lovers and believer sin animism. The sea and other natural elements are personified in their tradition and religion as they still live a sustainable way of life, just like their ancestors who never trawled, but instead utilized a unique way of fishing that does not devastate the coral or fish stocks. By portraying these two groups of people, the film establishes a dichotomy common in ecocritical debate: the opposition of the modern and the native, of development and conservation.

However, neither side of the dichotomy is adequate for facing the modern ecological predicament and the film thus presents the third group of people—researchers and activists, who act as mediators between the first two groups. Researchers bring indigenous wisdom into their scientific research and communicate the wisdom in a way that modern people can understand, while activists imagine new and sustainable ways of developing and living, and provide a third choice apart from the modern and the native. Here the term 'activists' differs from its usual meaning. These are not people who demonstrate and strike for environmental rights and policies, but there are people who try to incorporate social actions into their daily lives. They do not always possess the native understanding that the sea is alive, but acknowledge the importance of keeping the sea healthy. The film provides an example of Penghu residents restoring *Lao-Gu* stones—which are coral rocks utilized in ancient times to build houses—to the sea to prevent coral from further devastation. By engaging in such sustainable practices, they are in a broader sense activists.

In *Squid Daddy's Delivery Room* (hereafter *Delivery Room*) Ke documents the local Taiwanese environmental activist Guo Daoren, who builds an artificial, organic reef as a 'delivery room' for squids. Generally in Taiwan, artificial reefs are made out of old ships, and we are shown footage of ships blown up and sunk to the seabed, becoming artificial reefs. Inspired by what he witnessed in other East Asian countries, Guo decides to build a reef from bamboo, a place for squids to reproduce. The bamboo reef finally becomes not only a place for reproduction, but also a nodal point for marine engineers, environmental activists, and researchers to meet and exchange techniques, knowledge, and experiences. Thus the 'delivery room' not only facilitates the propagation of baby squids but also transforms the space into a new community construct for various species.

Elena Past points out in her analysis of Mediterranean cinema that underwater existence is 'essentially rhizomatic—members are connected through a complex system of non-hierarchical interrelations' (2009: 60). Footage of the interaction between divers and a school of fish in *Delivery Room* illustrates this rhizomatic existence as 'gravitational pull does not act on bodies in the same way, weights are lightened, and movement is elongated' (Past 2009: 57). As the fish swim in a circular motion and the diver slips into the circle, he reaches out to the fish as a friend or member of the same family. The narrator underlines this connection: 'the ocean for a long time has been degenerating due to human overfishing. But when you are in it, you could discern that the fish would not repel you. Rather, they would swim around you and you could join in anytime'. This intimate connection between humans and fish is described as a form of dancing: the fish sometimes swim around you and leave you alone at others, swimming in the same direction just like individual members of an orchestra playing a symphony. In the smooth space of the undersea world, human beings are no longer the descendants of *Homo erectus*. The elongated movement of the body turns human beings into another creature, or using Gilles Deleuze and Félix Guattari's term, 'becoming-animal' (2004). The journey of becoming-animal into the underwater world not only brings humans back to the primordial undersea ecosystem on the screen, but also makes possible the rebirth of life through the construction of the 'delivery room', which the environmentalists actively participate in. This environmental act does not leave the primordial ecosystem as such so that these divers can enjoy the return of the repressed nature; it rebuilds an ecological home for a new assemblage of human beings, bamboo reefs, fishes, squids and other marine lives.

'Eco-', the prefix of 'ecology', means habitat and environment as well as house or home and the home of the delivery room can be considered truly ecological as the reef is made of organic matter instead of iron or plastic. From this perspective, Guo and his assistant maritime engineer Cai are not only building a place for squids to propagate, but also creating an ecology for the new assemblage. The new ecology of the assemblage can be referred to as material ecology, which in general involves the entire physical universe. In this case, it is limited to the production of the bamboo reef, the reproduction cycle of squids and other marine life forms, and the material samples that go into laboratories for scientific research. Material ecology is perceived through technological ecology, which in *Delivery Room* includes undersea filming, lighting, narration, interviews and the soundtrack; that is, the technological relations through which the documentary is made. Undersea filming is the most important technique for mediating between the blue and the human world; it is the route 'through which we now can sense the world, most especially that part of the world's conversations which are not conducted in wavelengths we can hear, see or otherwise apprehend' (Cubitt 2005: 59). The material ecology, which could not be perceived without technological relations, is now made perceivable. The technological ecology brings a new aesthetics to the screen: the dominant colours of water in Chinese representations (yellow, azure and green) modulate into the liquid, light diffusing blue of the sea. This new aesthetics is not only sensational but ethical because the wonder or sublimity projected by the film can be a motivation for an ethical act, as is the case with Cai, inspired by Guo, undertaking his work, from which the third form of ecology is engendered.

The blue aesthetics forms the basis for the emergence of social ecology. For Guattari, social ecology

> [i]s imperative to confront capitalism's effects in the domain of mental ecology in everyday life: individual, domestic, material, neighbourly, creative or one's personal ethics. Social ecology will have to work towards rebuilding human relations at every level of the society. It should never lose sight of the fact that capitalist power has become delocalized and deterritorialized, both in extension, by extending its influence over the whole social, economic and cultural life of the planet, and in 'intension', by infiltrating the most unconscious subjective strata.

> (2000: 49–50)

The capitalist power in *Delivery Room* comes from the government and the fishing industry. When Guo suggests that the artificial reefs the government has built are harming the plant life in the coral reefs, a researcher from the fishing industry opposes his opinion by reiterating the age-old conflict between economic growth and environmental protection:

> [Y]ou say that we have the responsibility to save coral reef plants, but why do you jump to the conclusion that an artificial reef is harmful to the undersea environment? Don't you mean that you would rather let us starve and save the coral? It is unacceptable.

Guo, of course, understands the logic of the fishermen and how the protection of the coral reefs benefits the industry. However, what he wants to promote is much wider than the economic concern embodied by the researcher. He understands the interlinked relationship between the well-being of an ecosystem and economic prosperity. After all, both ecology and economics concern the knowledge of 'home' (eco). To him, the well-being of an ecosystem involves a new perception of nature and a new cultural praxis. By understanding the local ecosystem, a localized and reterritorialized social ecology can be conceived—a new assemblage of the sea that includes marine creatures, researchers and activists.

In short, Ke Chin-yuan's documentary provides the audience with a gleaming blue screen. Ke's 'blue screen' advocates that it is not enough to embrace a transregional vision of a resourceful earth, which is suggested by the green and colourless screens; a new ecosophy is required. The three ecologies in *Delivery Room* shift the silver screen from green to blue, a space in which human beings occupying the position of activists are represented as becoming-animal with life-engendering consciousness. Here, an inter-species assemblage is forged and an ecosophy combining the material, technological and the social is suggested.

Black Screen: Oscillation between Internationality and Postnationality

In contrast to the river imaginaries offered by mainland film-makers, which always halt at the borders and deltas, film-makers in Taiwan immerse their cameras into the sea, capturing the undersea world while at the same time explicating the intriguing transnational politics of protecting the ecosystems of the seas and realizing the transnationality evident in the flow of seawater. For example, *Du Dongsha/Travel to Dongsha Islands* (Ke Chin-yuan 2005) elucidates the intriguing transnational politics contained in environmental protection. The Dongsha Islands (or Pratas Islands) are governed by the government of Taiwan and are located in the north-eastern South China Sea, 340km south-east of Hong Kong, 440km south-west of Tainan and 850km south-west of Taipei. Fishermen from Taiwan, mainland China, Hong Kong, the Philippines and even Vietnam fish among these islands. In the film, Ke shows how mainland fishermen use poison and dynamite to harvest fish. These fishing methods violate Taiwan's laws, yet the army in Taiwan cannot enforce the law on the mainland fishermen as confrontation such as this would prove extremely sensitive politically. The army's only recourse is to issue them warnings and then to repatriate them back to China's maritime borders. The sea suffers not only because of the 'capitalist power' of the fishermen, whose sole concern is for more profit, but also because of the imposition of borders, which are, of course a human cultural and political construct, which divides the continuous landscape into different sectors under different sovereignties.

While the transnationality of Ke's *Travel to Dongsha Islands* (hereafter *Dongsha*) follows the conventional parameters for discussing Taiwan's transnationality, including the problems of nationhood and local identity, more complex evocations of transnationality is found in 'Kuroshio: Trilogy of Black Stream', consisting of *Heichao xinglü: zhuixun shenshi zhilü/ The Route of Black Stream: In Search of Identity* (Li Jinxing et al. 2003c), *Heichao sehngxi: taohairen de gushi/Living along Black Stream: Stories of Fishermen* (Li Jinxing et al. 2003b) and *Heichao maidong: haiyang wenhua de shengsi/The Pulse of Black Stream: Reflections on Ocean Culture* (Li Jinxing et al. 2003a). The Black Stream, widely known as the Kuroshio Current, is a north-flowing ocean current on the west side of the North Pacific Ocean that flows from the Philippines, past Taiwan and on to Japan. The Kuroshio Current is also called the Black Stream because the current appears black due to its lack of nutrient salts and absorption of sunlight. The smooth space of the sea now becomes striated. The striation brings sea creatures and nutrient salts from north to south and south to north. The striated sea is not the result of a human-imposed navigation. Rather, this is a natural phenomenon caused by the temperature differences of seawater. In this respect, the Black Stream can be referred to as a smooth space with a striated undercurrent. The 'black screen' created by the Black Stream offers another form of water imaginary and is a complement to the blue screen suggested by Ke Chin-yuan.

One of the more notable aspects of the black screen is Taiwan's aboriginal people, the *Yami*. In part 2 of the trilogy, *Living along Black Stream: Stories of Fishermen* (hereafter *Fishermen*), the film-maker attempts to trace the origins of the *Yami*. Searching for roots is not unfamiliar to Taiwan's cultural discourse. Back in the 1970s, the *Hsiang-t'u* (literally 'country and soil') movement started a debate between 'the West-leaning "modernists" and the hsiang-t'u nativists' (Yip 2004: 30). This was a debate of literary styles, resinicization and westernization, and later played an important role in the emergence of a 'Taiwan consciousness' (see Yip 2004). One of the key writers of this movement, Hwang Chu-ming, wrote a short story called 'Kanhai de rizi'/'Days of Watching the Sea' (1967), where the first half of the story portrays an entangled relationship between a young prostitute and *taohairen*, the colloquial way of addressing fishermen in Taiwan. When the young prostitute leaves the seashore and returns to the earth (*hsiang-t'u*) in the latter part of the story, the return signifies an identification with the soil (*t'u*) and a repudiation of the sea, which repeats an age-old cultural identity: a Han land-based nationalism/regionalism. About 30 years later, aboriginal cultures have gradually received more respect and a new ecoidentity has emerged. *Fishermen* (the title uses the term *taohairen* for fishermen) follows the Black Stream current and proposes that the *Yami*'s roots can be traced back to the northern Luzon in the Philippines. The people from Luzon and the *Yami* both belong to the Batanic linguistic group; their dialects are 80 per cent similar. This reverses the logic of Hwang Chu-ming, of 'leaving the seashore and returning to the earth', as it emphatically affirms that a sea culture is an inseparable part of Taiwan's cultural identity. Since the identity of the *Yami* is tied to the sea culture, their way of life is viewed as an oceanic ecological practice, which makes the identity not only a cultural but an ecological one (see Hu 2008: 45–107).

However, the explication of this sea culture and identity does not stop at the Taiwan–Luzon connection. *Fishermen* and part 3 of the trilogy, *The Pulse of Black Stream: Reflections on Ocean Culture*, extends the trajectory of cultural exchange along the Black Stream current by claiming that the three regions form a 'Black Stream cultural zone'. Hence, Taiwan and Japan connect not through the colonial past, but through a geoimaginary based on the natural phenomenon of the Black Stream and their shared environmental concerns along the current. Here, the so-called 'Black Stream culture' is primarily a fishing culture, consisting of the fishery industry (including fishing and aquaculture), the migration of fish and ecotourism within the region. Human beings from different sovereign areas and fish, along with other marine creatures, form a single ecocommunity. Their general well-being is interconnected and interdependent along the Black Stream. The demographic flow and common ecological concerns undermine the usual understanding of transnationality within Sinophone communities or Chinese-language areas. The transnationality in the trilogy goes beyond the problematic notion of ethnicity or a common-language community and proposes a transnational ecocommunity that connects Japan, Taiwan and the Philippines. The black screen reworks transnational power relations with ecological issues, while acknowledging that the sea is not an ideally smooth space in which the relationship between entities is non-hierarchical; rather it is a smooth space with a striated undercurrent on which a new transnational ecocommunity can be formed.

While the trilogy projects an ideal, transnational imagined ecocommunity that can think beyond national borders; this notion of ecocommunity is clearly too idealistic, especially when compared with *Dongsha*. The intricacies and complexities of international politics presented in *Dongsha* enforces the notion that the environment's well-being is dependent on international political relations. While the trilogy takes the audiences beyond the national borders to an ecological postnationality, to form a more realistic transnational ecocommunity, the political reality of the region should be put into consideration. In this vein, *Dongsha*'s political reality balances the far-fetched imagination of the trilogy. If there is anything *Dongsha* and the trilogy can offer to the discussion of transnational ecocinema, it is this oscillation between international and postnational modes of representation.

Conclusion

In this chapter, I have given a brief historical account of water imaginaries in documentaries produced in mainland China and Taiwan. In the mainland Chinese texts, *Yellow River* proposes a green screen with Chinese characteristics and represents water as an element of the flow of the river. It shows how the green and idyllic landscapes have been ruined by the yellow soil drained from the riverbanks and advocates that for the river to be 'green' again, humans need to fully utilize the energy of the river. This peculiar understanding of 'green' carries the philosophy of a socialist ecology: the ideal human–nature relationship is actually a nation–nature relation in which humans are pressing upon the limits of nature for national needs.

On the other hand, the eight-episode *Questions Concerning Water* offers a colourless screen with a particular representation of water, where it is no longer an element of a river but an isolated entity with its own identity. The 'colour' of the colourless appears in a water drop, which can be regarded as a quantifiable object that makes possible the concept of virtual water and a water-footprint. As China runs short on clean water, this colourless water becomes a valuable national resource. For *Questions*, a national water network is inevitable to channel water across the nation. The large-scale proposal of building the network irresistibly substitutes human actants for the national actant and turns the nation into a modern engineer who alters the natural landscapes according to its own needs and will. In this way, the green and colourless screens can be considered two faces of the same coin.

In Taiwan, Ke Chin-yuan's documentary captures the undersea world and provides the audience with a blue screen. The entire aesthetic and mechanics of existence of the undersea world differ utterly from the above-water world. To understand the aesthetics and mechanics of existence, a new ecosophy combining the material, technological and social aspects is required. The new ecosophy in *Delivery Room* captures a space in which 'activists' are represented as becoming-animal with life-engendering consciousness and an interspecies assemblage is forged. Ke's life-engendering consciousness makes his films truly ecological because the subject matter under his camera is the 'home' (eco-)—a place where different species encounter each other in a new assemblage.

Finally, the Black Stream trilogy creates a black screen, portraying the sea as a smooth space with a striated undercurrent that is the foundation of a new form of transnational thinking in Taiwanese cinema. Transnationalism in the trilogy differs from the usual cinematic discourse, which confines the scope of the discussion to Sinophone communities or Chinese-language films. Instead, it constructs a new line of transnationality by emphasizing the demographic flow and cultural exchange along the Black Stream from the Philippines to Taiwan and Japan. This black screen embraces a transnational cultural zone and is conscious of the common ecological concerns across national borders. The black in the trilogy is a line, linking up different geographies and ecosystems in different nations to form a broader ecological and transnational whole. Yet it is also a boundary that is not exempt from the complexities of international politics on environmental issues.

From the yellow to the green screen, from the blue to the black, this comparative study has described four kinds of human–nature relationships in place-based water imaginaries. By focusing on documentaries from China and Taiwan and the ways their arguments move from the national to the transnational, the variously coloured screens impart a sense of the ideological complexities of water-based cultural representation and ecocriticism in this era of ecological crisis.

References

Chen, X. (2002), *Occidentalism: A Theory of Counter-Discourse in Post-Mao China*, Lanham: Rowman & Littlefield.

China Central Television (CCTV) (1983), *Huashuo Changjiang/Discovering Changjiang*, China: CCTV.

—— (1986), *Huashuo yunhe/Discovering the Beijing-Hangzhou Grand Canal*, China: CCTV.

—— (1988a), *Heshang/River Elegy*, China: CCTV.

—— (1988b), *Huanghe/Yellow River*, China: CCTV.

—— (1997), *Sanxia beiwanglu/Memorandum of the Three Gorges Project*, China: CCTV.

—— (2006), *Zaishuo Changjiang/Rediscovering Changjiang*, China: CCTV.

—— (2008), *Shuiwen/Questions Concerning Water*, China: CCTV.

Chu, Y. (2007), *Chinese Documentaries: From Dogma to Polyphony*, London and New York: Routledge.

Connery, C. L. (2001), 'Ideologies of land and sea: Alfred Thyer Mahan, Carl Schmitt, and The Shaping of Blogal Myth Elements', *Boundary 2*, 28:2, pp. 173–201.

Cubitt, S. (2005), *EcoMedia*, Amsterdam and New York: Rodopi.

Cui, W. (2003), 'Zhongguo dalu duli zhizuo jilupian de shengzhang kongjian'/'The development of mainland Chinese independent documentary', *Twenty-First Century*, 77, pp. 84–94.

Deleuze, G. and Guattari, F. (2004), *A Thousand Plateaus: Capitalism and Schizophrenia*, London and New York: Continuum.

Feng, Y. (2007), *Bingai/Bingai*, China.

Gaga, J. (1999), *Color and Meaning: Art, Science, and Symbolism*, Berkeley and Los Angeles: University of California Press.

Gleick, P. H. (2009), 'China and Water', in Peter H. Gleick (ed.), *The World's Water 2008–2009: The Biennial Report on Freshwater Resources*, Washington, Covelo and London: Island Press, pp. 79–100.

Guattari, F. (2000), *The Three Ecologies*, London and New Brunswick, NJ: Athlone Press.

Hoekstra, A. Y. and Chapagain, A. K. (2008), *Globalization of Water: Sharing the Planet's Freshwater Resources*, Malden, MA: Blackwell.

Hu, Jie. (2006), *Chenmo de Nujiang/The Silence of Nujiang*, China: Green Earth Volunteers Office.

Hu, Jackson (2008), '"Spirits Fly Slow" (*pahapahad no anito*): traditional ecological knowledge and cultural revivalism in *Lan-Yu*', *Journal of Archaeology and Anthropology*, 69, pp. 45–107.

Ivakhiv, A. (2011), 'The anthrobiogeomorphic machine: stalking the zone of cinema', *Film-Philosophy*, 15: 1, pp. 118–39.

Ke, C.Y. (2003), *Amasi/Amorgos*, Taiwan: Taiwan Public Television.

—— (2004), *Jiyi Shanhu/Remembering Coral*, Taiwan: Taiwan Public Television.

—— (2005), *Du Dongsha/Travel to Dongsha Island*, Taiwan: Taiwan Public Television.

—— (2006), *Tianda dida/The Heaven and the Earth*, Taiwan: Taiwan Public Television.

—— (2007), *Chanfang/Squid Daddy's Delivery Room*, Taiwan: Taiwan Public Television.

Li, J., Qinhui, F. and Qinghui, Q. (2003a), *Heichao maidong: haiyang wenhua de shengsi/The Pulse of Black Stream: Reflections on Ocean Culture*, Taiwan: Taiwan Public Television.

—— (2003b), *Heichao sehngxi: taohairen de gushi/Living along Black Stream: Stories of Fishermen*, Taiwan: Taiwan Public Television.

—— (2003c), *Heichao xinglü: zhuixun shenshi zhilü/The Route of Black Stream: In Search of Identity*, Taiwan: Taiwan Public Television.

Li, Y. and Yu, Y. (2004), *Yanmo/Before the Flood*, China: Fan & Yu Documentary Studio.

Lu, S. H. (2009), 'Gorgeous Three Gorges at last sight: cinematic remembrance and the dialectic of modernization', in S. Lu and J. Mi (eds), *Chinese Ecocinema: In the Age of Environmental Challenge*, Hong Kong: Hong Kong University Press, pp. 39–56.

Lu, X. (2010), 'Rethinking China's New Documentary Movement: engagement with the social', in C. Berry, L. Xinyu and L. Rofel (eds), *The New Chinese Documentary Film Movement: For the Public Record*, Hong Kong: Hong Kong University Press, pp. 15–48.

Mi, J. (2005), 'The visual imagined communities: media state, virtual citizenship and television in *Heshang* (*River Elegy*)', *Quarterly Review of Film and Video*, 22: 4, pp. 327–40.

—— (2009), 'Framing ambient *unheimlich*: ecoggedon, ecological unconscious, and water pathology in new Chinese cinema', in S. Lu and J. Mi (eds), *Chinese Ecocinema: In the Age of Environmental Challenge*, Hong Kong: Hong Kong University Press, pp. 19–38.

Nichols, B. (2001), *Introduction to Documentary*, Bloomington, IN: Indiana University Press.

Past, E. (2009), 'Live aquatic: mediterranean cinema and an ethics of underwater existence', *Cinema Journal*, 48: 3, pp. 52–65.

Pinsent Masons (2011), *Pinsent Masons Water Yearbook 2010–2011*, London: Pinsent Masons.

Shi, L. (2004), *Nujiang zhisheng/The Voice of Nujiang*, China: Wild China Film.

Shi, Y. (2000), *Diansi jilupian: yishu, shoufa yu zhongwaiguanzhao/Television Documentary: Arts, Technique and Transnational Perspective*, Shanghai: Fudan University Press.

Wang, B. (2009), 'Of humans and nature in documentary: the logic of capital in *West of the Tracks and Blind Shaft*', in S. Lu and J. Mi (eds), *Chinese Ecocinema: In the Age of Environmental Challenge*, Hong Kong: Hong Kong University Press, pp. 157–69.

Wang, J. (1996), *High Culture Fever: Politics, Aesthetics, and Ideology in Deng's China*, Berkeley, Los Angeles and London: University of California Press.

World Bank (2006), *China: Water Quality Management—Policy and Institutional Considerations*, Washington, DC: World Bank.

Yan, D. D. (2010), 'Documenting Three Gorges migrants: gendered voices of (dis)placement and citizenship in *rediscovering the Yangtze River* and *Bingai*', *Women's Studies Quarterly*, 38: 1, pp. 27–47.

Yan, Y. (2008), *Yanmo 2: Gongtan/Before the Flood II: Gongtan*, China: Foggy City Studio.

Yip, J. (2004), *Envisioning Taiwan: Fiction, Cinema, and the Nation in the Cultural Imaginary*, Durham and London: Duke University Press.

Zhang, X. (2008), *Postsocialism and Cultural Politics: China in the Last Decade of the Twentieth Century*, Durham: Duke University Press.

Notes

1   This chapter uses the Chinese Pinyin for the names of rivers; i.e. Huang He and Changjiang rather than the Yellow River and Yangtze River.

2   TV documentaries in the course of PRC history were not always referred to as documentaries. In the 1980s, television programmes worked to distance themselves from those documentaries produced during the period of Maoist socialism, as the term 'documentary' was associated with a form of socialist propaganda linked to Maoism. Hence, the term *zhuanti pian* (special topics) emerged to differentiate the genre from *jilu pian*/documentary. Later in the 1990s, film-makers and critics, especially those from the New Documentary Movement, redefined the documentary film genre and readapted the term to contrast their productions with the *zhuanti pian*. This chapter uses the term 'documentary' to refer to both *zhuanti pian* and *jilu pian*.

From *My Fancy High Heels* to *Useless* Clothing: 'Interconnectedness' and Ecocritical Issues in Transnational Documentaries

Kiu-wai Chu

> For dancing girls of Chao-yang, token of profoundest favor,
> one set of spring robes worth a thousand in gold—
> to be stained in sweat, rouge-soiled, never worn again,
> dragged on the ground, trampled in mud—who is there to care?
> The *liao-ling* weaves takes time and toil,
> not to be compared to common *tseng* or *po*[1];
> thin threads endlessly plied, till the weaver's fingers ache;
> clack-clack the loom cries a thousand times but less than a foot is done.
> You singers and dancers of the Chao-yang Palace,
> could you see her weaving, you'd pity her too!
> (Extract from 'Liao-ling – Reflecting on the Toil of the Weaving Women', Po Chü-i)

In his poem '*Liao-ling*—Reflecting on the Toil of the Weaving Women', Tang poet Bai Juyi (aka Po Chü-I [1984] 2000: 772–846) relates the humble disposition of weaving women producing a high-quality silk (*Liao-ling*) robe for a lady in the Han Palace. He praises the intricacy and high quality of its production and the delicate beauty of the knitted work. The poem gives an account of the poor and disproportionate treatment of the dancewear by the emperor's beloved dance ladies who 'dragged [the dress] on the ground and trampled [it] in mud', indicating the ruling class' neglect for all the care and hard work endured by the weaving women. The poem's narrative reflects the power relations between the producer (the exploited) and the consumer (the exploiter) of China's feudal past. Although the poem was written over a thousand years ago, its two central issues—class inequality, high society extravagance and waste—are still prevalent in the contemporary consumer society.

Today, China has become the largest exporter of garments in the world. Each year, Chinese workers produce billions of pieces of clothing, and the number continues to grow with the speed of economic development. Yet, the garment industry continues to be exploitative, and factory workers in China still experience the same harshness as did Bai Juyi's weaving women. Separated by the passage of over a thousand years, China's economic development today is comparable to those of the world's most advanced nations and yet, changes in social environment of the clothes manufacturing industry appear minimal compared to feudal conditions. Social inequality and superfluous consumption of the rich remain the two major problems of clothes manufacturing. A notable change is that the world has become increasingly globalized where 'multinationals, far from leveling the global playing field with jobs and technology for all, are in the process of mining the planet's poorest back country for unimaginable profits' (Klein 2001: xvii). Class inequality is no longer just a national but a global issue, which involves not just particular groups but everyone in the world.

While Bai's poem was written at a time when environmentalism as defined today was not in existence, a present day reading of it as an ecocritical text suggests that concern for social and environmental values existed since ancient time. Parallels can be drawn with recent transnational documentaries[2] and their critiques of the contemporary global garment industry in terms of global social inequality and environmental awareness. With reference to Jia Zhangke's *Wuyong/Useless* (2007), and Ho Chao-ti's *Wo Ai Gaogenxie/My Fancy High Heels* (*MFHH*) (2010) for comparison, this chapter explores how transnationalism in these films, on one hand, facilitates the expression of ecocritical and environmentalist messages concerning nature and the environment, as well as the relationships between human and all other beings on the planet. On the other, transnationalism reveals the challenges and contradictions film-makers and artists face in productions attempting to reflect ecological and environmental concerns, challenges and contradictions that are brought about by the unstoppable forces of excessive global consumerism.

Fashion, Clothes and Ecology in Transnational Documentaries

In one of his most acclaimed photo series' 'Manufacturing' and his documentary *Manufactured Landscapes* Canadian photographer Edward Burtynsky powerfully represents the essence of the daily activities of Chinese factory workers in sets of squarely framed prints. The workers, who spend as many as 15 to 18 hours a day in the factory, are shown fixated at their operating place, hunched over sewing machines, immersed in their mechanical and repetitive work. To them, work is their whole life, and the factory, their entire world. Other than in the works of Burtynsky, such scenes are also commonly seen in transnational documentaries, as in *China Blue* (2005), *Useless* and *MFHH*, films and other media concerning China's modern industrialization. Recently, transnational documentaries that convey a strong message of increased interconnectedness between nations have been highly popular. Released in 2005, Canadian film-maker Micha X. Peled's *China Blue* has caught particular attention in the Western world. Following a thread cutter in Shanxi, a teenage girl named Jasmine from distant Sichuan Province, the film reveals the harsh life experiences and exploitation of

Figure 1: Manufacturing #15. Bird Mobile, Ningbo, Zhejiang Provinnce, China, 2005, Edward Burtynsky.

factory workers. In one scene, we see images of exhausted factory workers sleeping between heaps of newly produced jeans, while in another, we are shown female workers keeping themselves awake by using clothing pins on their eyelids, so that they may take on additional quotas to produce more jeans for consumers in Western cities. *China Blue* has established itself as a model transnational documentary, with a mission to denounce entrepreneurs and large international corporations, while fighting for equality and welfare of workers in developing countries.

Such transnational documentaries[3] are increasingly committed to issues of global environmental justice, as well as revealing exploitative profit-driven production practices, with the assumption that transnationality in film enables us to see their global connectedness more readily. These documentary works reveal the growing cross-regional communication and exchange that entail the emergence of transnational culture that is no longer anchored in a specific place. Such filmic representations facilitate us to make sense of the global network of products and flow of messages, and make us aware of the physical and symbolic linkages between those seemingly disconnected worlds. From the food we eat and the news we read, the jeans and shoes we wear; to the world films, music and performances we enjoy, our living habits and daily routines today are unimaginable without global networks of information and exchange, regardless of which part of the world one is situated in. In that context, reference to *MFHH* and *Useless* suggests how transnational documentaries represent the interconnectedness that, to some extent, generates a sense of ecological awareness, which has emerged as a major theme in recent Chinese documentaries.

Figure 2: *Useless*, China Film Association, Mixmind Art and Design Company, Xstream Pictures. Workers in a garment factory in Guangzhou.

MFHH: Shoes, Labour, Cows

[M]aking a pair of shoes can be really complicated. It involves obtaining and using materials from many different places. At the time I thought, if I traced further and further upstream from this production line, what would I see? And where would my search for these people lead me?

(Ho 2011)

Triggered by this thinking, Taiwanese film-maker Ho Chao-ti's documentary *MFHH* portrays in detail the production line of a branded pair of high-heel shoes, and shows us how the making of such a commercial product reflects the impact of globalization, and how countries and places become connected as a result of their social and economic interests and demands.

Ho shows these through her effective use of graphical images of the contributive parts of the production process. Visits to where these take place introduce us to people of different social strata whose lives are associated with high-heel shoes. The film begins with an interview with a middle class New York designer, a fashion fanatic who owns 52 pairs of branded high-heel shoes. To her, each pair is a unique art piece. In her words, 'it's a form of sculpture. It's a language'. The film then shifts to a high-heels factory in Guangdong in Southern China, where the shoes are seen manufactured into finished products for packaging and export. Next, further back in the production process, viewers are taken to a leather plant in Northern China to see how raw leather is industrially treated for manufacturing, and finally to a slaughter house close to the border with Russia, where baby calves are slaughtered and their skins collected. The film explores downstream along the production line to show, behind the scene, fundamental problems of very large income gap between the haves and the have-nots, the different wealth situations between rich and poor nations and regions of the world. From high-end affluence to extreme poverty, one sees disparities behind the facade of 'global interconnectedness'.

While such transnational documentaries draw attention to social and class inequalities, such as focusing on how poor working conditions in the factories are causing workers higher exposure to health and technological risks, *MFHH* takes one step further from its concern for human subjects to greater socioenvironmental justice towards the non-human world. While the film basically addresses problems of economic globalization, our attention is drawn also to its ecocritical portrayal of the slaughter of baby calves and, with it, the exploitative practice of global fashion production in the contemporary world.

MFHH opens with two tied-up baby calves being dragged into a slaughter house, signalling the imminent fate of these calves. Following scenes of interviews in the shoe factory

Figure 3: *My Fancy High Heels*, Conjuncture Films. A butcher slaughters a hanging cow.

conducted with production line managers and workers, viewers are led to discover that the many large sheets of lifeless grayish black material for making shoes are in fact pieces of calve skin once stained with tissues of their organs and blood. Viewers begin to see the connection between the lifeless shoes one wears and the living cattle somewhere else in the world. The film takes us further to slaughter houses in Northern China where we see how live cows are hung up shivering in pain and in fear with their empty and helpless looks before their heads and limbs are chopped off and their blood drained. Viewers are then shown how butchers skillfully slice off their skin on the blood-drained floor. Here, baby calves are normally killed within 12 hours of their birth, so as to obtain best quality young skin that can be made into high-grade leather for production of brand-named high-heel shoes.

In closing, *MFHH* shows us images of two extreme worlds, with one scene inside a slaughter house showing high overhead piles of fresh-looking cow leather still dripping in blood and fat, and another of a fashion show in progress somewhere in a Western metropolitan city with parading fashion models showing off their brand-named high-heel shoes. The film closes with an animated scene of a baby cow peacefully lying beside its mother.

Figure 4: *My Fancy High Heels*, Conjuncture Films. Close-up shot of a cow with an empty and helpless look.

As forceful criticism of the cruel practices involved in the production of high-heel shoes, *MFHH* reveals the transregional processes of how a pair of leather shoes are transformed from its original form as a living animal. In other words, its focus on this process reveals in reverse how an animate is killed and converted into an inanimate.

Mane points out that

> Nature is silent in our culture ... in the sense that the status of being a speaking subject is jealously guarded as an exclusively human prerogative [...] For human societies of all kinds, moral consideration seems to fall only within a circle of speakers in communication with one another.
>
> (1996: 15–16)

In *MFHH*, the representation of the process of reanimating the inanimate thus returns to nature (in the form of natural living such as animals) its power to speak, which facilitates viewers to reassess critically the adverse impacts of modern industrial practices.

Such attempt contributes towards developing an ecocinema, or in Ursula Heise's term, a cinema of 'ecocosmopolitanism', which 'reaches towards what some environmental writers and philosophers have called the "more-than-human world"—the realm of non-human species—but also that of connectedness with both animate and inanimate networks of influence and exchange' (Heise 2008: 60–61). Heise is concerned with the question of 'how the rights (or more generally, the affectedness) of non-human parts of the biosphere should be legally, politically, and culturally represented' (2008: 61). Eventually, an ecocosmopolitan cinema transcends national and cultural boundaries, and envisions 'individuals and groups as part of planetary "imagined communities" of both human and nonhuman kinds' (Heise 2008: 61).

Useless: Clothes, Memories, Taste

While Peled's *China Blue* and Ho's *MFHH* sharply criticize the garment industry's inequitable production process and exploitation of factory workers in China, Jia Zhangke's tripartite documentary *Useless*, which also deals with the process of garment production and consumption, offers more obscure points of view towards the negative impacts of global commercial activity. By portraying how different people communicate with the world through the way they make, use, wear and repair clothes, *Useless* connects people of various places and social backgrounds, and leaves viewers to interpret for themselves any message that Jia might have intended in the film.

Since his earlier works such as *Xiao Wu* (1997), *Platform* (2000) and *Unknown Pleasures* (2002), Jia Zhangke's concern has always been for the poor and marginalized working class in backward Chinese villages, particularly in remote towns in his native province of Shanxi. In time, as he became more well known, his interest and vision gradually extended to areas beyond Shanxi. Tony Rayns has rightly pointed out that the emphasis in Jia's films since *The World* (2004) has changed 'from using settings in less known locations of Shanxi and focusing on narratives of insignificant individuals or small communities, to emphasis on bigger impact stories, with multiple characters and greater settings' (Jia 2009: 231). On this view, Jia's growing interest in placing different fragments in a film narrative has coincided with his desire to portray more groups, or communities, of people spanning different localities. In *Dong* (2006), through portrayal of the artist Liu Xiaodong, the Three Gorges in China and Bangkok in Thailand, two distant settings in Asia, are connected. In his view,

> [I]t is difficult to show the complicated and multi-faceted human endeavors that one feels in real life, if we tell a story of just one couple using traditional and enclosed settings within a 90-minute narrative. In today's world, it is common to interact with people of unrelated interests and develop relations in and between different spatial settings.
>
> (Jia 2009: 231)[4]

And, in *Useless*, through the theme of clothing, Jia portrays the interconnectedness between three geographically distinct locations, Guangdong, Paris and Shanxi, especially in relation to the ways production, consumption and appreciation of clothes connects them.

Despite being less provocative than *China Blue* or *MFHH* when it comes to labour exploitation, mass production line operation, and adverse impacts of globalization, *Useless* is forthright in its call for humanitarian values in the production of manufactured goods. In a garment factory in Southern China, Jia took numerous shots of workers on an assembly-line that show the mechanical and inhumane working conditions. Concerning the conditions that make the production of global commodity possible, Walter Benjamin once stated that '[a]s a consequence of the manufacture of products as commodities for the market, people become less and less aware of the conditions of their production' ([1938] 1983: 104). Mechanical mass production severs relations between the 'makers' and 'users'. Because of that, one can easily replace an old piece of goods with a new one. But by so doing, one creates wastage in resources and deprives oneself of memory. Thus memories and affects for objects can no longer be formed. Such phenomenon echoes the realities of China's modernization today, where the desire for anything new means throwing away all things old.

After documenting the mechanical daily routines of garment factory workers, Jia interviews a renowned Chinese fashion designer Ma Ke and shares her philosophy on fashion and life. Despite being a successful designer with her own commercial brand, named 'EXCEPTION', Ma Ke is highly critical of seeing clothing merely as commodities. To her, the major distinction between artwork and commodity lies in the reasons why they are created: '[t]hose created for faith and joy are art; those for fame and benefits are commodities' (Tsui 2010: 186). Ma's views have led her to create an alternative line of *haute couture* that is based on a non-assembly-line mode of production. She calls it 'Wuyong' (Useless), due to its lack of practical use.

To explain her intention of creating the 'Wuyong' series, Ma Ke cited Tang poet Meng Jiao's 'Wanderer's Song':

The thread in the hand of a kind mother
Is the coat on a wanderer's back.
Before he left she stitched it close
In secret fear that he would be slow to return.
Who will say that the inch of grass in his heart
Is gratitude enough for all the sunshine of Spring?[5]

Meng's poem suggests that in former times, the making of a coat required the use of needle-and-thread by hand and the finished product was filled with human sentiments. The coat made by the kind mother is therefore particularly treasured by her son on a distant journey.

With the long, laborious process of production, Ma stresses that clothes, or hand-made objects in general, convey memories, emotions and affections to which mass-produced objects can never be compared. To respond to the loss of such human sentiments in mass-produced clothing in the contemporary world, Ma Ke developed her idea of the 'Wuyong' series by burying a number of her designer dresses underground for 2 years, letting nature take its course on them. When the clothes were dug up, they would be imbued with the time and place they were buried, with traces of their histories, carrying with them memories that were shaped by nature.

The soiled shirts and dresses were dug out in 2007 and exhibited at the Lycee Stanislas during the Paris Fashion Week. They were worn and displayed by models of all ages, selected from a wide range of French people, from professional models, dancers to street artists, who are used to performing as living statues for hours in the open. Through using these Chinese-soil 'natured' clothing to discredit 'memory-lacking' mass-produced products, her fashion series is presented as a form of ecological, counter-consumerist art performance. The Useless line has exceeded the practical function of clothing, to convey spiritual messages about the value of memories and psychological feelings towards the importance of time. Ma Ke has created this non-commercial fashion series to question the need for the extensive speed of China's rapid development, and offer criticism on the exploitation of the underprivileged, social inequality as well as excessive waste resulting from the development.

However, Jia's film does not end with Ma Ke, as during the third part of *Useless*, he takes us from Paris to a backward village in Shanxi, China. As Ma Ke notes in the film, the aim of creating her 'Wuyong' series was precisely to enable her to be closer to people far away from the cities, to include peasants and villagers in the highlands and mountain areas. She suggests that when one goes to such natural and simplistic places, and sees how people live, one has a better chance of recovering lost memories and will gradually remember things already forgotten. In the remote mining areas of Shanxi, Jia follows a couple of villagers and miners socializing in small local tailor shops. In one of these shops, a shabbily dressed elderly

Figure 5: *Useless*, China Film Association, Mixmind Art and Design Company, Xstream Pictures. Ma Ke's 'Wuyong' exhibition, Paris Fashion Week.

villager enters with a pair of torn old trousers for repair, a common habit of old villagers in Shanxi who have their old worn clothes repaired again and again. The film continues to relate slices of local life by telling the story of a happy coal miner couple. The husband, formerly a tailor, left his old job because of the decline in the tailoring industry. He recalls his purchase of a pink suit for his wife and genuinely praises his wife's beauty in whatever clothes she wears. The wife, in return, expresses how much she treasures the three-piece suit because it was chosen by the husband. In another scene, Jia shows a young man on a motorbike chasing another rider, giving us the impression that they are in a competition. The young rider rides stands on his bike, the top half of his body bared, enthusiastically waving his flag-like white shirt, as he is energized further by the rallying cries, aptly demonstrating characteristics of his youthful impatience. And, at the end of the film, viewers are taken to another tailor shop to witness an ordinary encounter between an old woman and a tailor. From their conversation, we learn about the villagers' concern with the appearance of large garment factories in their area, the imminent decline of their small businesses, and their uncertain eventuality. Through his portrayal of the characteristics of different age groups amidst the Shanxi villagers—young, adult and old—Jia shows us how they treasure the clothes they wear and how memories and human feelings are invested in 'clothing'. The film contrasts the villagers' clothes that are invested with human values with mass-produced garments that are without personal sentiments and memories. What the latter has is only their monetary value, a transient product of consumerism.

Figure 6: *Useless*, China Film Association, Mixmind Art and Design Company, Xstream Pictures, A local knitter in Fenyang, Shanxi.

According to Jia,

> *Fashion* has become the most used word in China, with the *new rich* clamouring for goods of branded names like LV, Armani, Prada... Behind this consumerist passion, wealth has become the most important value judgment or the norm for personal standing in society.
>
> (Jia 2009: 233, original emphasis)

In the global capitalist world today, branded names are closely associated with 'taste' which, in Benjamin's view, 'develops with the definite preponderance of commodity production over any other kind of production. [...] [A]s the expertness of a customer declines, the importance of his tastes increases—both for him and for the manufacturer' ([1938] 1983: 104–05). According to Benjamin, what is considered taste today goes hand in hand with ignorance.

With transnational documentaries like *MFHH* and *Useless*, clothesmaking processes in different parts of the world are revealed, the unseen linkages and global interconnectedness between makers and users are represented and made visible. As spectators, we are made aware of the exploitative practices of the processes in factory production, and are therefore no longer consumers ignorant of these processes. Can one's 'taste' for commoditized clothes then be subverted or destroyed, leading people to adopt more ecological attitudes towards the clothes they wear? In *Useless*, Jia uses Ma Ke's work as a case study to express the complexity of such issues, and the conflicts between consumer culture and environmentalism.

As discussed earlier, what Ho Chao-ti's *MFHH* attempts to achieve is to return the speaking power to 'nature'. Through the chain of connections with baby calves, the pair of voiceless high-heel shoes are used to narrate their story of how they transformed from a lively animate to a silent inanimate. In much the same way, Jia's *Useless* leads us to contemplate whether, with Ma Ke's act of burying clothes underground for a period of time, clothes that originally conveyed no memory or human sentiments can be reinvested with missing qualities and an ability to tell their own stories given to them by nature, through the medium of film.

With nature's presence appropriated in the soiled clothes, however, this chapter questions whether such investment of meanings in clothes truly reflects a world where different parties are globally interconnected, or instead, do they expose the inadequacies and reluctance of garment industries and the current fashion scene in bringing about ecocritical rhetoric?

Rethinking Ecocritical Issues in *Useless*

Transnational films, according to Lucia Nagib, should be seen as a polycentric space where 'everything can be put on the world cinema map on an equal footing' (2011: 1). However, whether such 'global interconnectedness' reflects real situations of cross-regional communication and represents everything and everyone 'on equal footing' remains hugely debatable. Behind the harmonious representations of transnational communications, is 'global interconnectedness' merely a disguise for the hegemonic structure and its adjacent regional disparities, which portrays transnationality with a series of ideological assumptions? Or do transnational documentaries like *Useless* and *MFHH* reveal to us the complexity and ambivalence in clothes production and consumption in contemporary societies, which leads film-makers and viewers to engage in a kind of ecocritical reflection upon themselves?

By representing the concept of 'global interconnectedness' and portraying cross-regional communications among people in different socioeconomic contexts, *Useless* sets out to reveal the discrepancies between the worlds of the rich and the poor, and to provoke viewers to reflect on the varied meanings of 'nature' to different communities, so as to re-evaluate contemporary societies' extravagant fashion and consumer culture. At the same time, by promoting Ma Ke's meaningful and righteous philosophy in ecofashion, *Useless* appears to reassert the problems arising from the global capitalist system that the film sets out to criticize. Through his objective filmic approach towards the subjects (the garment factory workers, Ma Ke and her fashion series, to the villagers in Shanxi), Jia presents Ma Ke's ecocritical thinking behind her creations, but at the same time, the difficult and contradictory situation she is in as both a commercial branded fashion designer and an ecoartist, necessitates Jia to engage in self-reflection and self-criticism of his role as a film-maker and consumer.

Dividedness within Interconnectedness

From Ma Ke's 'Wuyong' fashion exhibition in Paris to the remote village in Shanxi province, Jia Zhangke continues to the transition of locations via the film narrative. In the closure of the second part of *Useless*, Jia shifts his lens from the dimmed, atmospheric interior of the exhibition hall at the Lycee Stanislas to long shots of a wasted mining land in billowing sand dust, beyond which only factory constructions and a motorbike riding pass can be seen faintly in the distance. Jia takes us to his home village in Fenyang in China. In the shots that follow, Ma Ke is seen driving her Japanese car in the countryside. As the car passes through the woods, she relates the inspiration she obtained from these remote places away from the cities, places that are hard to get to, including remote villages and nature sites such as mountain and plateax areas, which inspired her to create the 'Wuyong' series. In Ma Ke's final scene in the film, we see her driving past a local villager in Fenyang. Jia's camera stops following her, and instead follows the elderly villager to a local tailor shop.

Apart from the sudden shift of setting from Paris to Shanxi, what we also see is an interesting shift in our identification from the renowned fashion designer to the villagers of Fenyang. As Ma Ke drives away from the camera, we as spectators are made to stay with Shanxi villagers for the rest of the film. Putting Southern China, Paris and Shanxi in the same film context, *Useless* appears to have placed people of different nationalities, social strata and backgrounds, who share similar sentiments toward certain societal activities, on the same platform. However, with the shifting identifications from Ma Ke to the Shanxi villagers, can we then conclude that they are represented as equal subjects?

In *Useless*, Jia adopts two types of documentary representation—the observational and the interview-oriented[6] style—nimbly to represent different subject characters, depending on a person's social background and lifestyle in different parts of the world. Although observational and interview-oriented documentaries have their distinct differences, they both aim to present objective interviewee accounts by letting them speak their factual experiences, the former by capturing the subjects' daily life trivia without intervention, and the latter by letting the subjects voice out their personal accounts. Jia's employment of both the observational approach of film-making as well as interviews of different characters in different contexts is particularly apparent in the representations of the diverse settings of the film. In the garment factory, the film is constructed of emotionally detached observations of the ordinary life of factory workers. Numerous shots capture their immersion in the work, yet we are not provided any interview with the workers. With Ma Ke, however, Jia changes his technique to monologue-like interviews in which Ma explains her ideas of creative design and her critical views on assembly line methods of garment production. In Fenyang, Shanxi, Jia adopts a more balanced approach by carefully observing the daily life routines of old town folks, on one hand, and, occasionally, popping a question or two to villagers he was filming, reminding us of his directorial presence behind the camera.

Zhang Yingjin states that in recent Chinese documentaries, conducting interviews has become a common practice, as a way to offset official viewpoints with individual ones. With the use of a collective view obtained from conducting interviews with a number of individuals, it is possible to break down an authoritative discourse. Through testimonies the subjects made in front of the camera, a sense of authenticity is enhanced in the documentary.[7]

(2006: 57)

However, it is worth noting that film-making of any kind entails a process of selection: from what and who to film; when to conduct an interview; when to employ techniques of pure observational filming; to, at the end, which recorded parts of the film to use or discard. These are all parts of the process that could influence the interviewee's subjectivity and in turn, reveal not the subject's but the film-maker's point of view. While Jia is well known for his fascinating use of narratives with humanistic concerns, in *Useless*, such humanism is only partially seen, particularly in the section shot in his home village in Shanxi. Unlike Peled's *China Blue*, Jia's prime focus is not on how factory workers in Guangdong are being maltreated or exploited, but on how assembly line production severs relations and feelings between people.

With evenly allocated screen time among the three parts of the film, *Useless* keeps the oppressed disempowered. By doing so, the film subtly reveals the environmental injustice felt by certain communities in the contemporary world. Jia has said that through *Useless*, the three groups of people have been brought together, connected, so to speak. However, behind the facade of that linkage, Jia leads us to reassess Ma Ke's creations in ecocritical terms, by revealing the hierarchical, imbalanced power relationships among different groups. In other words, the imbalanced use of interviews among different characters in the film, in some ways, gives them different rights in voicing their conditioned subjectivity. While a renowned fashion designer is able to relate her creative ideas and philosophies toward life in great detail, villagers in Shanxi are given less chance to speak their minds and the Guangdong assembly line workers are not allowed to voice themselves. Likewise in *MFHH*, being the characters in the lowest strata in society, butchers in the slaughter house are not given interview opportunities. The absence of voice becomes a kind of disempowerment to certain groups of people, reasserting their status as the marginalized and voiceless poor. In *Useless*, through giving different characters different degrees of speaking power, the film-maker has, to a certain extent, deprived viewers of the agency of the workers, leaving them as characters that are voiceless, mechanical and marginalized, a method use for criticism of the dehumanizing clothing industry.

From Ma Ke's 'Ecofashion' to Jia Zhangke's 'Ecocinema'

According to Jia, 'everyone is interconnected, whether they be tailors of Shanxi, or workers of clothing assembly lines in Guangzhou. They are all parts of a chained network, despite their professions or hierarchy in society' (Gao 2008: 72). However, *Useless* emphasizes the contrasting styles of living, and thus the dividedness among different social groups, which in turn reveals their different perspectives towards clothing and the environment, as well as the meaning of nature. It is the juxtaposition of 'interconnectedness', and 'dividedness' among different social groups, which allows Jia to invite multiple judgments towards Ma Ke's 'Wuyong' series.

In *Useless*, Ma Ke remarks, 'When you go to remote, natural places and see how people live, it seems like recovering lost memories. You gradually start to remember things you once felt'. She suggests that people from different socioeconomic backgrounds are interconnected by the possession of shared lost memories for 'nature'. It is such memories that bring us all together, that make all livings 'globally interconnected'. In creating her 'Wuyong' series, Ma stresses that in order to instill a sense of 'memory' into the clothing dug out from the ground, she had to be strict in her selection of suitable soil for treatment of the clothes in her studio. In doing so, the unearthed clothes will resemble, and be on par with, the dust and dirt-stained clothes worn by poor peasants and miners. 'Soil' in the film, is then given the symbolic meaning of nature's power of healing. Through her work, Ma attempts to express the importance of returning to nature, in order to recover the contemporary city dwellers' lost memories and sense of feeling. To put theory into practice, she built her studio in Zhuhai, Guangdong, in an area surrounded by lots of greenery, a place where even her pet dogs can wander freely. Jia sees Ma's decision to move the core of her work and personal life to the countryside as a reflection of her ecocritical standpoint (Jia 2009: 234–35). With the idea of creating an ecofashion series, Ma Ke's 'Wuyong' attempts to question the validity of China's speed of development, the reasons for such speed and development, its obliteration of memories and its excessive wastage of natural resources. And, by showing the transnational links among various parts of the world in relation to fashion and clothes, Jia's film enables viewers to see its imbalanced social realities, so that one could ponder questions of memory, consumerism, the decline of local trades and industries, social injustice faced by the poor, as well as environmental damages, from a more global perspective.

Despite this, Jia's *Useless* is not a one-sided propaganda film that unreservedly supports Ma's philosophy. The garment factory shown in the first segment of the film is clearly the place where EXCEPTION, Ma Ke's commercial line of clothing, is produced. Although Ma does not directly disavow her earlier commercial venture in the interview, Jia's questioning puts her in an unavoidably self-contradictory and possibly embarrassed position. On one hand, her brand EXCEPTION has brought her fortune and fame, while, on another, she rejects mass production of clothing in the market in favour of her new non-commercial product 'Wuyong'. In *Useless*, Jia gives his complex and ambivalent response to the validity of Ma Ke's ideologically contradictive works, which are the end products resulted of an excessively developed consumer society in China today. When seen from this angle, we can also reinterpret Jia's distance from the factory workers he filmed as he made no attempt to enter the inner world of their minds, neither to understand their social lives after work. Instead, he quietly observes their assembly line working behaviour to show them merely as mechanical, silenced, dehumanized and objectified workers. This is, after all, how they are treated as part of the modern industrial practices.

Tony Rayns points out that Jia has 'dedicated particularly lengthy sequences to depicting the relatively less developed parts of China, focusing on revealing the living conditions and health issues of factory workers and village miners' (Jia 2009: 233). By placing together sequences of people from different socioeconomic backgrounds, the juxtapositions bring into question the validity of Ma Ke's creation, as they question whether her attempt in 'soiling', thus 'naturalizing', the clothes for the 'Wuyong' series necessarily conveys a universal, ecocritical message. Unlike *MFHH*, where people of different countries are found on the same economic chain and production process, the subjects in *Useless*, such as the wealthy Parisian fashion community and poverty-stricken Shanxi villagers, do not engage in any physical contact or communication, but are symbolically connected only by the two types of look-alike clothing, both being soiled, torn and tatty. Commenting on the parallel between Parisian fashion models' mud-coloured face and body make-ups, and the dust filled bodies of Shanxi miners in a public bath, Jia states in an interview, '[i]n France, dusty mud is for the face. In Shanxi, it gets sucked into people's lungs' (Gao 2008: 72). One could argue that these 'nature-enhanced' clothes are merely objects for ideological investment where, by appropriating primitive, 'natural' third-world aesthetics in first-world conceptual fashion, transnational imagined communities are formed, where different worlds appear to become connected, or even united.

Figure 7: *Useless*, China Film Association, Mixmind Art and Design Company, Xstream Pictures. Dusty clothes of Shanxi miners.

Eventually, Jia makes no attempt to provide a simple conclusion. Instead of reaffirming the status of 'Wuyong' as an ecological fashion project, Jia is more interested in examining Ma Ke's idea in producing the series, so as to provoke viewers to reflect on the different understandings of nature by individuals from various socioeconomic contexts. Ma Ke's intention in creating a series of non-commercial fashion is to remedy the lack of concerns for human values by the assembly line mode of industrial production and consumerism, in support of designs that are simplistic, natural and ecological. But Jia Zhangke's ambiguous stand in the film has rendered her creative work somewhat self-contradictory. Viewers are generally aware that Ma's resolve to venture into creating her non-commercial 'Wuyong' series, comes from the fame and fortune she has earned in her commercial line EXCEPTION that, to this day, continues as a brand-name in China. Through *Useless*, Jia expresses his admiration of and respect for Ma's unique idea and design of her creation, which nevertheless provides effective marketing for Ma's commercial products. On the other hand, Jia also expresses his scepticism towards assembly line mode of production that makes his film both ironic and ambivalent. Instead of making direct comments or criticism on this contradiction, Jia cleverly uses Ma's own words to bring out her views on such contradictions prevailing amongst the drastic socioeconomic developments in contemporary China. Simultaneously, the film invites us, viewers and consumers, to reflect on our often extravagant and wasteful ways of consumption and the material excess, which are often enabled at the expense of the exploitative industrial practices

Figure 8: *Useless*, China Film Association, Mixmind Art and Design Company, Xstream Pictures. Parisian fashion model in mud-colored make-up.

and social injustice experienced by the underprivileged, in the increasingly transnational world today.

Conclusion

The film image impresses us with its completeness, partly because of its precise rendering of detail, but even more because it represents a continuum of reality that extends beyond the edges of the frame and which therefore, paradoxically, seems not to be excluded. A few images create a world.

(Macdougall 1998: 132)

One could indeed state that in transnational cinema, the world *is* connected and created with a few images. From Ho Chao-ti's *MFHH* to Jia Zhangke's *Useless*, representations reveal the contrasting conditions of different places and people. Recent transnational documentaries, which emphasize 'global interconnectedness', respect for and return to nature, have brought people of different economic and social levels together and represent them in world cinema as more equal beings living in a cosmopolitan world. To a great extent, transnational documentaries enable global spectators to be aware of issues of environmental injustice and exploitation faced by the underprivileged in the world.

While *MFHH* is a focused and effective transnational environmental film that exposes the exploitation and inequality generated in the complicated process of production of branded high heels, Jia's *Useless* shifts the emphasis from criticizing the exploitative assembly line productions to the loss of human values, memory and affect in mass-produced goods, as it promotes fashion designer Ma Ke's design and creation philosophy. Jia presents Ma's 'Wuyong' fashion series that reflects her intention in reinvesting humanist values and an appreciation for nature into clothes, and leads viewers to reassess the idea of 'ecofashion'. One could argue that by promoting Ma Ke's philosophy, Jia has inevitably promoted her reputation and commercial products that, to a certain extent, reaffirms the exploitative global capitalist system other ecodocumentaries aim to criticize. However, by juxtaposing how diverse groups of people perceive clothing and the environment differently *Useless* stresses the discrepancies and dividedness among different groups and reveals the socioenvironmental injustice experienced by the underprivileged.

In transnational films, when narratives take on a cross-regional context, the geographic specificity of a 'place' often becomes more emphasized. Yet, film-makers have attempted to identify and represent commonalities, such as the dusty clothes worn by Parisian models and Shanxi villagers in Jia's *Useless*, to claim a sense of global interconnectedness between different locales. But transnationality does not necessarily imply an elimination of national or regional boundaries. The actual interconnectedness does not lie in the aesthetic commonality of different places, rather their actual 'differences' or discrepancies. Transnational cinema will convey ecocritical messages effectively only if we retain our awareness of the differences among the different worlds, and reflect upon ways of eliminating them in order to bring justice and equality to different groups in the increasingly connected world.

References

Benjamin, W. ([1938] 1983), 'Addendum to "The Paris of the Second Empire in Baudelaire"', in Rolf Tiedemann (ed.), *Charles Baudelaire: A Lyric Poet in the Era of High Capitalism* (trans. H. Zohn), London: Verso, pp. 103–06.

Berry, C. and Farquhar, M. (2006), *China on Screen: Cinema and Nation*, Hong Kong: Hong Kong University Press.

Gao, R. (2008), 'Jia Zhangke yu Wuyong de mimi', *Chinese Textile and Apparel/Zhongguo Fangzhi*, June, 2008, pp.70–72.

Graham, A. C. (trans.) (1968), 'Meng Chiao's wanderer's song', in A. C. Graham (ed.), *Poems from the Late T'ang*, Harmondsworth: Penguin Books, p. 63

Heise, U. K. (2006), 'The Hitchhiker's guide to ecocriticism', *PMLA*, 121: 2, pp. 503–16.

——— (2008), *Sense of Place and Sense of Planet: The Environmental Imagination of the Global*, Oxford, New York: Oxford University Press.

Ho, Chao-ti (2010), *Wo Ai Gaogenxie/My Fancy High Heels*, 56 minutes, Taiwan. Conjunction Films.

——— (2011), 'Director's statement', *My Fancy High Heels*, http://myfancyhighheels-en.blogspot.com/2010/09/directors-statement.html. Accessed October 15, 2011.

Jia, Z. (2007), *Wuyong/Useless*, 80 minutes, China/France. China Film Association, Mixmind Art and Design Company, Xstream Pictures.

——— (2009), *Jiaxiang 1996–2008: Jia Zhangke dianying shouji*, China: Peking University Press.

Klein, N. (2001), *No Logo*, London: Flamingo.

Macdougall, D. (1998), *Transcultural Cinema*, USA: Princeton University Press.

Manes, C. (1996), 'Nature and Silence', in C. Glotfelty and H. Fromm (eds), *The Ecocriticism Reader: Landmarks in Literary Ecology*, Athens and London: The University of Georgia Press, pp. 15–29.

Nagib, L. (2011), *World Cinema and the Ethics of Realism*, London: Continuum.

Nichols, B. (1985), 'The voice of documentary', in Bill Nichols (ed.), *Movies and Methods, Vol.II*, USA: University of California Press, pp. 258–73.

Po, C. (aka Bai, Juyi) ([1984] 2000), 'Liao-ling—reflecting on the toil of the weaving women', in Watson Burton (ed.), *The Columbia Book of Chinese Poetry: From Early Times to the Thirteenth Century* (trans. Burton Watson), New York: Columbia University Press, pp. 245–46.

Taylor, L. (1998), 'Introduction', in David Macdougall (ed.), *Transcultural Cinema*, USA: Princeton University Press, pp. 3–21.

Tsui, C. (2010) *China Fashion: Conversations with Designers*, Oxford and New York: Berg, pp. 164–88.

Zhang, Y. (2006), 'Style, subjects and special point of view: a study of contemporary Chinese independent documentary', in Jie Ping (ed.), *Reel China: A New Look at Contemporary Chinese Documentary*, Shanghai: Wenhui Chubanshe, pp. 53–68.

Notes

1   *Liao-ling* (繚綾), *tseng* (繒) and *po*(帛) are different kinds of fine Chinese silks. Among them, *liao-ling* is the finest of all.

2   This chapter takes 'transnational' to refer 'to phenomena that exceed the boundaries of any single national territory' (Berry and Farquhar 2006: 5) to indicate international co-production and the film's production and representations of various countries or regions, and in line with the view of transnational cinema 'not as a higher order, but as a larger arena connecting differences' (Berry and Farquhar 2006: 5) between nations.

3   To name a few, recent transnational ecodocumentaries include *China Blue* (Micha X. Peled, 2005), *Manufactured Landscapes* (Jennifer Baichwal, 2006), *Maquilapolis* (Vicky Funari and Sergio de la Torre, 2006), *Dong* (Jia Zhangke, 2006), *Chinatown* (Lucy Reven, 2009), *Waste Land* (Lucy Walker, 2010) and *Cotton Mill* (Zhou Hao, in production).

4   The quote is translated by this chapter's author from the original Chinese text.

5   This version of the poem is translated by A. C. Graham, in *Poems of the Late T'ang* (1965: 63).

6   Observational cinema is a mode of documentary developed in the early 1960s, when new cinematic technologies enabled documentarians to film individuals and interactions in various settings that had not previously been possible, resulting in a shift in emphasis from the public to the private, and from the general to the particular (Taylor 1998: 4). In the 1970s, interview-oriented documentary films gradually burgeoned, offering interviewees the opportunity to face the camera and have direct dialogues with the audience. This makes the audience witness-participants of interviewees' life accounts, and offers them the ease to identify better with the interviewees (Nichols 1985: 259–60).

7   The quote is translated by this chapter's author from the original Chinese text.

Ecocinema and 'Good Life' in Latin America

Roberto Forns-Broggi

Introduction: Ecological Imagination and Ecocinema

E cological imagination is a key concept for understanding the connections and systems of the natural world. Because of the monopoly of imagination in the media to the detriment of other uses of imagination, ecological imagination can be hard to explain. To start, an ecological imagination depends on both conscious and unconscious mechanisms of control in our dominant rationality. Ecocritical cinema has enormous potential for the investigation of the ecological imagination. According to Greg Garrard, recent ecocritical work on cinema has been scarce despite the far wider cultural currency of visual media as compared to print (2010: 18). Certainly, a wider interest in reflecting on films and new media will be encouraged by the growing number of active viewers who form parts of the environmentalist organizations from all over the world. Several communal organizations and other alternative media circuits are capable of producing and circulating all kinds of films and videos that commonly are not available through the mainstream channels. In particular, it is worth calling attention to an unusual website, *Culture Unplugged*. This is an interactive online film festival, where film-makers, producers and distributors from different continents screen interesting films and tell stories on indigenous rights, social conflicts, spiritual practices and other current topics. On this site, it is easy to find films on current ecological conflicts like the political disputes regarding the extractive industries (mining and fossil fuels) and biomass conflicts (deforestation). Indeed, there is a wide range of film choices for discussing contemporary directions in ecocinema.

To effectively select films that deal with environmental conflicts, one must start with a clear and functional definition of ecocinema. One can articulate the following definition based on the first ecomedia seminar of the Association for Studies of Literature & Environment (ASLE 2011): Ecocinema is a classification for films that lend themselves to generating ecological awareness, which reflects a consciousness about both fruitful and problematic relations with natural life. However, this ecological awareness does not have to be explicit or intentional. Also, films that mask or eliminate that awareness should be considered part of ecocinema studies. It is the imperative of an ecocritic to study films and videos, regardless of any superficial 'environmental' value, for their ecocinematic potential (Monani 2011).

Yet, the field of ecocinema remains contested territory. Adrian Ivakhiv concludes in his panoramic article, 'Green Film Criticism and Its Futures', that '[t]here is much work to be undertaken by ecocritics interested in unpacking the environmental meanings carried or enabled by, and the constraints and potentials inherent in, film and visual media' (2010: 23). Ecocinema is not a finished idea. Neither is it a prescriptive form of environmental media. Even though it is a concept under construction operating with and from different local traditions, environments, economies, and territories, ecocinema works to assemble film festivals and academic studies with the intention of challenging our habits of perceiving this 'more-than-human world'. If we consider ecocinema as a seedbed for cultivating an ecological imagination across boundaries of races and nations, we must also profess awareness that this notion of ecocinema is not easy to describe. Not only because it is a notion under construction, but also because it depends on diverse cultures and environmental factors.

Ecocritics working on Latin American films and new media need to ask questions such as: How does ecocinema encourage viewers to become more ecologically minded observers? How does ecocinema help viewers to become more sensitive to the ways in which films reflect, shape, reinforce, and challenge our perceptions of nature? Ecocritics need to study and discuss ecocinematic theories and practices; they need to figure out how ecocinema engages environmental justice, narratives of risk, and critique of globalization. Indeed, it is necessary to have a better understanding of the social and cultural realities fostered by the cultural dominance of technologies and lifestyles modeled on hegemonic power. These social and cultural realities continue to create audiences, who remain immersed in colonialist dimensions of complicity and apathy. This also distances the audience from a sense of immediate responsibility, proposing a role reliant on mere information and instruction. Ecocinematic spectatorship needs to cultivate the responsibility to produce, to create, and to give back. If the audience learns to respond to ecocinema in a practical way and to reach beyond the immediate reality of their daily lives, then ecofilms can truly start to matter. Following from this notion of learning, when producing ecocinema about sustainable living practices, the film-maker needs to make explicit the connection between local reality and the knowledge and experiences to be derived from other cultures and places.

Filming the Good Life

Ecocinema in Latin America does not only study and produce films and media as tools for raising awareness and educating people about environmental issues. It is also a platform for the cultural conception of 'Good Life' or *Buen Vivir*. This key concept, emanating from indigenous ways of communal life between the human and non-human world, was recognized in the new Constitutions of Ecuador (2008) and Bolivia (2009). Those constitutional incorporations reflect a significant change in the deep-rooted marginalization of indigenous people in Latin American countries. A precursor to Good Life is the notion of 'conviviality' that anthropologists used two decades ago to highlight the affective sociality of indigenous communities (Overing and Passes 2000: 16–17).

Although Good Life shares similarities with the concept of conviviality, it remains a relatively new concept in the current political debates. This constantly under construction concept attempts to rethink conventional philosophies emphasizing economic development as it is based on conception of society in which human beings mutually support each other and are in harmony with nature. Many diverse backgrounds encourage and nurture this plural notion, from intellectual reflection to civil involvement, from indigenous traditions to alternative academies (Gudynas and Acosta 2011: 71). Some seekers of Good Life, according to Eduardo Gudynas, are not only indigenous communities and organizations, but they also consist of Western thinkers from diverse disciplines, including critical studies of development, biocentric environmentalism, radical feminism, postcolonial epistemologies, among others (2011).

Good Life has several names: In Spanish, *Vivir Bien*; in Aymara, *Suma Qamaña*; in Guaraní, *Ñanda Reko*. In Quechua, the major language of the Andean region, Good Life is called *Sumaq Kawsay*. In Bolivia the concept remains *Sumaq Kawsay*, but other languages vary the words and the meaning of this key concept as, for example, in the Mapuche culture in Chile and Argentina, *Küme Mongen* means 'living well in harmony'. Gudynas underlines this diversity of meanings as a way to germinate an alternative to the dominant idea of economic development. He also points out that this notion resembles certain concepts of deep ecology:

> All of them, suma qamaña, ñande reko, sumak kawsay, deep ecology and many others, complement each other, show some equivalence, converging sensibilities, and it is precisely this complementarity which allows a delineation of the building space of Good Living (2011).

While each form of expression has a different emphasis and accent, all of them refer to 'a way out' of the materialistic dominant culture and a way of life based on paying respect to the environment and people. For Gudynas, Good Life 'appears as the most important current of reflection which Latin America has offered in the past few years' (2011). To underline its break with conventional economic thinking, this concept remains important as it relies on indigenous traditions and visions of the Cosmos; it breaks with modern concepts of development and focuses on humanity's relationship with nature (Fatheuer 2011: 17). Good Life is different from consumerist well-being. It is a notion in opposition to present economic growth, because it is a reaction to this development's limitations and contradictions (Gudynas and Acosta 2011: 81). This notion responds to old problems, such as overcoming poverty and reaching equality, and to new problems, such as loss of biodiversity and climate change. It offers a new way to collectively construct different lifestyles and alternatives to lifestyles based on material progress (2011: 81).

Latin American ecocinema can offer a remarkable contribution to elaborate the importance of the concept of Good Life. More than a space for environmental representation, 'cinema' is, in Adrian Ivakhiv's words, about movement, extension, and the ongoing process by which meaning and affect are generated out of this movement (2011: 123–24). Considering the complexity and vastness of ecocritical studies, it is important to choose the type of cinema one wants to see and create: an art that offers a way of taking a profound look at things, a kind of aid to people in delving deeper and discovering new meanings of life. How ecocritical studies engage with this type of cinema is precisely something that we will keep evolving in the near future, both as a promising field of transnational studies and as a useful way to provoke creative reflection. Ecocinema can help us to understand the concept of Good Life, to illustrate it, to document it, and to establish channels of communication between different cultures in moving toward a sense of planetary interconnection. Thus, it is of vital importance that Latin American communities embrace this challenging platform to create productive dialogues and increasingly convincing activist cinema.

Sometimes Latin American ecocinema overlaps with indigenous alternative media, although the two classifications do not always completely coincide. If one were to access the vast selection of films and videos available at festivals, art houses, libraries, and at times from the film-makers themselves, one might come to the assertion, perhaps a generalization, that the ecocinema of the Americas (North, Central and South) tends to focus on rural life threatened by the process of modernization. The indigenous film-makers, and also Latin American and foreign artists analyse, contemplate, and celebrate the rural areas; and denounce the threats and ruinous polluting effects of modernization. These film-makers explore and focus on a lifestyle that relates to traditional forms of culture and which can be considered to fall under the concept of Good Life. This alternative, complex, and communal style of life where community includes nature can be very difficult to grasp for a filmgoer that pursues a more pleasant and relaxed film experience.

Even though the notion of Good Life is well known in Latin America, as Gudynas and Acosta observe, it is practically omitted from official directions by the economically dominant hegemonic forces in the Americas. The prevalent idea of economic 'well-being,' on the other hand, has been internalized and naturalized as a vital component of economic growth, and by extension, social life. The concept of Good Life is a very promising alternative to these dominant conceptions of progress, but as Francisco Salgado observes, it remains an unfinished and uncertain idea, especially when it comes to its ability to provide a functional and practical social model:

It remains to be seen whether *Sumaq Kawsay* contributes to the social construction of a new and just nation or the Quechua symbolic way of living together stays as just a notion.

The question is: Can *Sumaq Kawsay* really be a new model that allows a new order based on increased and empowered social participation—deliberative citizenship—that would lead to actions from both the state and the markets and nurture life in all its forms?

(Salgado 2010: 206)

Ecocinema of Good Life: Frames for Political Documentaries

In the emerging field of ecocinema studies, the notion of Good Life enables us to explore dialogues between Andean and other communities as they strive to imagine alternatives to the dominant consumerist lifestyle that is being promoted and enforced by the current hegemony of economic thinking. Before embarking on such a project, it is necessary to avoid compartmentalizing the diverse forms of production which, among other things, seek to awaken not only an ecological consciousness but also a commitment to act ecologically in ways that crosses frontiers and goes beyond labels. A first step in this direction will be to approach this diversity of documentaries and fictional films as articulations of Good Life, and from thereon build a common platform based on the practice of interculturalism that focuses on the harmonious relationship between human beings and the natural world.

The political documentaries *When Clouds Clear* (Danielle Bernstein and Anne Slick 2008) and *When The Land Cries: Operation Devil* (Stephanie Boyd 2010) form part of the ecocinema that can be considered as based on transnational collaborative initiatives. These films are greatly influened by the notion of Good Life. Both films belong to a practice of collective film-making that Sophia McClennen describes as 'a collaborative practice whose process is organically tied to its product' (McClennen 2008). In these documentaries the local people appear as agents of social transformation and a model of political resistance precisely because they illustrate a practical example of Good Life. As Belinda Smaill describes in *The Documentary: Politics, Emotion, Culture*, this type of collaborative practice forms a mutual social agenda among film-makers, technicians, and main actors which surpasses empathy to initiate a recognition of the other in its specificity through a dissident gaze that arises from the civic capacity to question the base of the political state (Smaill 2010: 91–92).

This dissident gaze, evoking a level of political consciousness, is present in *When Clouds Clear* and *When Land Cries: Operation Devil*. In that sense, the supportive practice of film-making and of paying attention to the most representative spokesmen of the community, represents a position that shows the importance of civic love and loyalty when forming a collective, a quality recognizable in documentaries of the dissident gaze. Examples of such documentaries include the meditative films of Abbas Kiorastami; two documentaries on the political-cultural struggle of the Mapuche people: *Uxüf Xipay/The Plunder* (Dauno Tótoro 2004), and *La voz mapuche/The Voice of the Mapuche* (Pablo Fernández and Andrea Henríquez 2010); and the disarming activist creativity of The Yes Men. Also, the tradition of media criticism 'that perceives the medium as a complex of power-political instruments' includes

Guy Debord, the situationists, and also works by Jean-Luc Godard, Pierre Bourdieu, Vilém Flusser, and Perter Sloterdijk, among others (Vit Havréanek 2008: 142). Each one articulates an agenda related to this genealogy of dissident documentary, although the mechanisms of validation and distribution of this alternative discourse are still very precarious.

Latin American ecocinema offers such a dissident gaze by assembling the shared components of Good Life in diverse ways. First of all, ecocinema offers an ethic imperative to recognize and assign importance to nature and people. It expresses traditional knowledge and indigenous practices as well as fosters calls to create unions, dialogue and interaction between different epistemologies of knowledge. Ecocinema introduces extended communities in which the river or the mountain are active members, like the spirits who are present in the Amazon tribes and cities. All these components of Good Life in the ecofilms serve to reinforce the importance of an active and activist response and resistance to neoliberal practical rationality, which has been enforced for decades in the name of progress and economic growth. What is more, this neoliberal rationality denies its role as the main cause of the pollution and plunder of the ecological resources of the rural communities of Latin America. Ecocinema fights common assumptions about ecological crisis by deliverling images that verify the seriousness of the crisis and by showing that the components affected are indeed proof that our society has an entirely weak self-consciousness. Furthermore, they are also proof of a self-destruction which is more extensive, slow, and devastating, against which it offers powerful images of solidarity as encouraging manifestos for Good Life.

When Clouds Clear

If one seeks a film with powerful images of solidarity, *When Clouds Clear* would be a good choice. Produced by American documentarists Danielle Bernstein and Anne Slick with the assistance of the people of the rural community of Junín in Ecuador, the documentary is about a conflict with corporations from Japan and Canada who were attempting to extract natural recourses from this community. The film records the division that occurs between the locals who are resistant to the economic incentives offered by the corporations and those who accept the terms offered. The film emphasizes the testimonies of the resisters, each one with their own perspective, who are all committed to the value of Good Life—which eventually feeds to their collective courage in defense of the land. This sense of collective is present early on as the beginning of the film introduceds us to the importance of collective work they call *Minga* labour. For the individuals which form the small community of resistance, Minga is the only value that makes sense. Indeed, to these people Good Life means living their daily life in solidarity: 'It has always been a custom here, to lend a hand', states one of the residents.

The documentary unmasks the manipulated mass media version of the Junín population's struggle against the displacement enforced by the multinational mining corporations. The confrontation with corporate power is presented through a Manichaean lens in the headlines of newspapers and in television interviews with official spokesmen. The portrayal of the conflict is obviously biased against the position of the people of Junín who are resistant to selling their lands. The official Ecuadorian newspapers speak of the subalternized features of the population of Junín, classified as figures of extreme poverty. The rhetoric of 'progress', persuasive and convenient for a practical rationality, employs its usual scientific and civilized tone in the media; headlines often use such terms as 'terrorist' and 'violence'. The media spokesmen appear logical and speak as 'experts'. They give a categorical image of political correctness and they project the idea to the general public that to be opposed to extractive activity is without justification and even comprises anti-national vandalism. Some images portrayed in the documentary are controversial, especially those of the mining facilities that were burnt down by the resisters. The only coverage of these events in the official media focuses on the 'brutality' of opposition activities against the mining corporations.

While the media only assumes the corporate point of view, *When Clouds Clear* tries to delineate the portrayal of the resisters as 'invisible' by the media, creating a contrast to the 'visible', 'civilizing' position enjoyed by the spokesmen of the mining corporations. The documentary offers a political view that challenges the hegemonic media representations, calling for national progress as it shows how the Manichaean version does not really do justice to the ecological beliefs of the resisters. While the film highlights their bravery, their solidarity, and the reasoning behind their political actions, the vigilant gaze of the documentary is not doe-eyed, but rather complex as it avoids an idealized portrait of the resisters. Far from presenting an inflexible representation of their political objectives for justice, the documentary explores the multifaceted testimonies of the protagonists. However, this does not mean that the film does not show the courage of the resisters, courage that is made evident in the testimonies and the defensive action of the people. This is how we see their desire to preserve the area of Junín and to resist the offering of money and properties, goods and benefits from the mining corporation.

Absent-minded spectators might simply reject the documentary, distancing themselves from the 'ecologist' rhetoric and perhaps from the demonization of the mining businesses. The primary focus of the resisters' protests, however, is more affirmative of the explicit values of love for the land, solidarity with the community, and the unity of families with the land, forming one entity, which will not accept the exploitation and division proposed by activities of the mining corporation.

The components of Good Life intertwine with the social commitment inherent to Minga labour to imply a clear rejection of the mining companies' offers. The interviews show the dignity of the political action of the community and the consciousness of its power to resist the aggressions of the corporations, despite the lack of governmental support. The resisters appear in the documentary as agents of social transformation and as a model of political resistance that defends Good Life, which appears as a way to engender their positive emotions towards the endangered landscape. What we end up with is a collage of discourses and perspectives, some Manichaean, others more complex. The documentary also leaves us with the impression of a variety of emotional expressions: desires and motivations, the contrast of the worries and goals of Good Life. On the one hand, the love, which the land expresses, integrates the local people into itself. What is at stake here is a way of life, Good Life. On the other hand, the collective thought process of the resisters contrasts itself with the pursuit of money (which often comes with threats and bribes) and with the profusion of bureaucratic speechifying that manifests its shameless character by destroying everyday lives. One dramatic, yet revealing scene shows paramilitary soldiers—called 'guards' and described as having been 'kidnapped' by the newspapers—peacefully disarmed by resisters and taken hostage in the chapel of the town. The documentary shows an alternative side to the official story as it piles on one testimony after another expressing family divisions and consequent distancing; we follow the irrevocably painful words and facial expressions of the resisters who, at the end of the film, appear silent in black and white images.

When Clouds Clear evidences something that perhaps the spectator does not want to admit. That like the people of Junín who accepted the incentives of the mining company, we accept the home that the market has made for us, the sophisticated and exaggerated consumerism, the ethics of corporate work, and the daily routine of the remunerated work in the urban context. The subjects, who resist being inserted into the reductive image of those who dedicate themselves exclusively to work and consumption, also resist the transformation of their lives into a consumed and desolate landscape.

A Dissident Gaze on Freedom, Equality, and Solidarity

The most impressive aspect of *When Clouds Clear* is its explicit resistance, ardently defended: the boy Robinson Piedra who speaks of the fight and of his education; the leader Polivio Pérez who not only shows off his verbal eloquence and great intelligence in the management of critical situations, avoiding violence, but also his uninhibited and admirable civic love which does not allow itself to be threatened or swayed by offerings of corporations. The eloquent young woman Marcia Ramírez and her aunt who fears nothing and bravely faces the threatening paramilitary soldiers; and Mr. Rafael Piedra, one of the founders of Junín, who appears throughout the documentary next to his wife Carmen Berrones. All of them, portrayed with the lively and relaxed gaze, surrounded by their nearby relatives, do not just represent the resistance to the power of money, but they *are* an embodiment of that very same resistance.

When Clouds Clear is not part of a political campaign, but this does not mean it has no power to support a political side of the portrayed conflict. Leftist critics often excessively stress the importance of representing the resistance, but documentaries of dissident gaze seek to be self-existent as a gaze, not conforming to the opinions of the film-maker. It is a gaze that we can adopt. It is a gaze that sees itself as a gaze of concern for others, as a new comprehensive rhythm, as a disturbing movement of reality. The dissident gaze does not end with the movement of the camera. It demands a vigilant conscience, it probes deep into us, and into all the ways we conspire against what we imagine to be our most urgent needs.

The assembly of testimonies becomes a movement that reflects the movement of the same clouds that fleetingly occupy some images in the documentary. Good Life is a living concept that looks to the future. These images leave us at the end of the documentary with a feeling of uncertainty. The resisters are able to see themselves through their own dissident gaze and through the eyes of the media as well. We do not see the local people who went to work for the corporation. After the fog, what is visible? What can be thought if the blockages that solidify our sense of separation and indifference toward each other are removed? What can be seen *When Clouds Clear*? Is Good Life still there?

The Clandestine War for Natural Resources

The conflict of indigenous peoples with mining companies all over the world is not a new phenomenon and it is worth mentioning several videos campaigning against this on YouTube. Mining is a main cause of pollution and depletion of natural habitats that are essential for entire communities. These communities have often been completely abandoned by their national governments. While Good Life is defended by popular organizations, State power is presented in mainstream media as the prime position to defend, ignoring the rights of nature. For Fatheuer, the exploitation of nature and mineral resources is central to the question of whether *Buen Vivir* will be able to gain political traction (2011: 26), and these are some of the issues which are addressed by these grassroots documentaries.

One film that successfully portrays the complex issue of open-pit mining is *Tierra sublevada: Parte 1, Oro impuro/Land in Revolt: Part One, Impure Gold* (Fernando Solanas 2009). It recounts a case of depletion of natural resources in the context of an economy designed to benefit a minority of power groups, most of them multinational corporations operating without major restrictions in Central America, South America and Africa. A short documentary *Laguna negra/Black Lagoon* (Michael Watts 2009) focuses on the social effects of the mining process in an actual rural community in the north of Piura, Perú, close to the region where the events of *When the Land Cries* took place. It describes the case of the torture of 30 rural dwellers by police and private security forces.

Watching both Peruvian documentaries, an obvious connection emerges: the mining companies hired the same security company, Forza, which enjoy impunity from the crimes committed against the leaders and members of the indigenous organizations that defend their environment. These films probe a crucial aspect of environmental destruction: it is caused by corporate terrorism.

But this is not a simple crime since it is extremely difficult to accuse these powerful companies who constantly change their names. For example, in Watts' short film, Monterrico Metals S.A. was owned by the huge Majaz mine when abuses occurred. Later on this mining company was renamed Río Blanco Copper S.A. In 2007, it was sold to the Chinese corporation Xiamen Zyin Tong Guan, which is owned in part by the Chinese government. The slippery profusion of identities allows corporations to evade legal liability, which is something that grassroots and online filmic documentation of abuses aims to correct.

When the Land Cries

In *When the Land Cries*, Stephanie Boyd chooses the format of an espionage thriller to approach the entanglement of economic interests and influences behind the anxieties and threats that the priest Marco Arana and his closest supporters suffered. Boyd's documentary deals with a counter espionage operation in the clandestine war for natural resources. Through this, the film provides a specific setting in which to discuss the rights of nature that the Ecuadorian constitution has recognized as *Sumaq Kawsay*. The priest Marco Arana, who can be considered a defender of Good Life principles, discovers that ex-secret service agents of the Fujimori regime, who nicknamed him 'Devil,' are following him. This priest has been a key mediator between the state forces and the indigenous organizations in Cajamarca trying to stop mining companies from opening new sites of extraction. Arana embodies a solid argument that denounces the plunder of natural resources against Good Life in *When the Land Cries*. He was also a founder and director of a NGO, Grupo de Formación e Intervención para el Desarrollo Sostenible/Group of Formation and Intervention for the Sustainable Development (GRUFIDES) and is an active member of Red Latinoamericana de Conflictos Mineros/Latin American Network of Mining Conflicts. He has published an article about natural resources as merchandise (Arana 2008: 11–15), and is thus a powerful embodiment for the fight against the exploiters of Good Life.

Through Arana, Boyd reveals a corrupted system of pressures and threats above the law, and through this the audience can get an understanding of how economic power works, and how important these social movements are for the defense of their rights and the rights of the land—both of which converge in the notion of Good Life. This is something that Freya Schiwy identifies in the indigenous videos: fictional narratives and Hollywood styles are adapted to Andean storytelling, symbols, and textile technologies (2009: 14–15). The storylines of this thriller closely follow the suspense of an action film, but it also shows a deep connection to what the priest and the other farmers and protesters refer to as the needs of the community: the defense of the land and its sacred elements, especially the water which is needed in very large quantities for the mining process. But this is not the only element that plays a key role in the protests against mining companies.

Anthropologist Marisol de la Cadena studies the enormous change in politics that is occurring in the Andes with the inclusion of mountains as political actors (see also Li 2009). The usual way of making political decisions has changed by the inclusion of cultures that for centuries were excluded from the political arena. De la Cadena mentions the same 'sacred place', the Cerro Quilish, which the priest Arana defended among protesters in Cajamarca from the Gold mine Yanacocha. In Boyd's documentary, even though the mountain is not portrayed as a political actor, it clearly remains a key subtext in the discursive defense of human and environmental rights (De la Cadena 2010: 357–62). This defense challenges the idea held by corporations and the state that geographical space on the periphery is remote, unproductive or even empty. For the resisters, open-pit mining does not allow any personal relationship with the mountains. The mountains are destroyed, erased, and the barren soil is left without any trace of life.

Another key asset in publicizing the efforts of the indigenous communities and figures like Arana is the exhibition of these films at festivals. GRUFIDES in coordination with the Denver Justice & Peace Committee and the Denver Film Society organized a screening and discussion of the film in Denver in October 2010. Thanks to the public protests against open-pit mining all over the world, especially in the countries of the Global South, we can note an increase in the visibility and awareness of social and environmental impacts of mining. Since the Denver screening and discussion about *When The Land Cries,* more links, maps, and updated information were noticed on websites such as No a la Mina and Defensaterritorios among others. More informed spectators will hopefully think of creative ways to take action after seeing an ecofilm of this kind instead of merely becoming depressed by the facts of pollution and devastation caused by mining companies.

Short Films and Experimental Videos about Good Life

An idea confirmed in the production of movies and alternative videos by native artists and groups is that the new media can transform people into designers of meaning in participatory processes. The proliferation and increasing number of active multimedia online venues is very promising for ecoart in Latin America because it creates a space to articulate new perspectives on ecological thinking. As a way of offering an introduction to this sort of visual production I will reflect on a couple of visual jungle poems based on the biodiversity of Machu Picchu that explore the possibility of video art focused on new ecological perspectives, emotional states of ecological consciousness, and the pursuing and sensing of connectedness and relationships between humanity and nature.

Jungle Tattoo (Forns-Broggi 2009) is a 5-minute visualization of a poem about being a part of the environmental history of Machu Picchu. Here I wanted to explore the possibility of visualizing one key component of the notion of Good Life: an extended community can be recognized in the ancient Incan ruins from the point of view of native birds. *The Bird's Dream* (Forns-Broggi 2011) is another 5-minute short film with the insertion of shots of birds to suggest a dreamlike sequence at a New Jersey shore and at The Ballestas Islands, Paracas. The short is an attempt to show the extended community of diverse birds via the artistic platform of a dream, where the montage of images calls attention to the life of birds and emphasizes the idea of nature's agency. One example of this idea is the consideration of bird songs as a sentient response by beings earnestly engaged in the present moment. In the context of an iconic place for tourism and national identification—the ruins of Machu Picchu—these little Andean dwellers are ignored or overlooked. By drawing out the birds from this iconic setting, the spectator is encouraged to question the usual ways of perceiving relating to the natural environment.

Other short films that are available online, such as *La Tierra es nuestra/The Land is Ours* (Carlos Alvarez Zambelli 2010), continue this ecological project as they will certainly prompt interesting discussions on the notion of Good Life. The aforementioned short is a 4-minute film that incorporates urban animation with clips from other documentaries about the indigenous struggles for survival against pesticides and other polluting effects of modernization in Latin America. The interesting feature in this video is its global perspective as it predominantly focuses on the interconnectedness of European and American cities, and in the process revealing from where many natural resources originate. This short film is part of an environmental campaign of the Valencia branch of an international non-profit organization, Association for Cooperation in the South (ACSUR). Other videos express environmental deterioration more symbolically. For example, *El llanto de la tierra/The Cry of The Earth* (Lucio Olmos Morales 2008), a 7-minute short film available on YouTube which compares the process of extraction of natural resources to the suffering body of a dying woman. Good Life is here equated to the sadness of Gaia as the personification of Earth as a self-regulating giant organism capable of adjusting to constant changes, small and great. Clearly this tale of diminishing Good Life implies the need to rethink extractivist practices. As Gudynas observes, 'because of their social and environmental impacts, extraction activities are clearly incompatible with Good Life in all its concrete expressions' (2011).

Why pay so much attention to the increasing number of farmers who have been organizing 'Environmental Defense Committees'? It is obvious that they pursue Good Life. It is important to want, as they do, to stop big corporations, such as mining companies supported by the Peruvian government with media coverage and open corporate support. Part of this venture is finding new points in which ecological perspectives can expand. Instead of creating media projects as an isolated effort to raise ecological concerns about humanity's separation from natural environments, the notion of Good Life demands a daily compromise of connections and actions within communities in the process of film-making. While some of my video projects are similarly explicitly ecological in content, they did not reach their intended audience because the production process did not occur in collaboration with local communities. One step to overcome this lack of community dialogue is to embark on a greater variety of methods to media production. My own experience in making a very low-budget short film about books that were burnt at the National Library in Sarajevo in 1992, *The House of Wisdom* (Forns-Broggi, 2006), taught me the benefits of digital technology, especially its very low cost. The lack of the collective dimension of my projects does not stop my questions about these attempts to capture and translate into images the notion of Good Life.

Another challenge to improve the circulation and resonance of ecocinematic projects in the context of Andean countries is to use regular media channels and to create an alternative place to promote a dialogue that includes communities of active viewers able to generate more complex responses. These videos could transform the community of viewers into art practitioners, provoking different types of responses: such as project publicity, ecoart installations, and international community involvement in land and water remediation. It is difficult to create a space that allows the embracing of the multiplicity of Good Life and the overcoming of the territorial confines of human subjectivity. Ideally, these videos would test the power of the ecoart project to create a platform for dialogue about Good Life that responds to the needs of the particular historical background and biodiversity of each community.

This need to meditate on this issue of Good Life responds to an attempt to cultivate a kind of global environmentality. That is, a mentality that thinks and acts in effective ways to resist the destructive practices of our societies on order to cultivate gentle connections to where we live, to re-establish time/spaces of solidarity, and communal consciousness, among the other mentioned components of Good Life. One should attempt to build into these ecoart projects perspectives that are more than just upgraded versions of animistic Good Life. If one wants to reach communities beyond local borders, one must formulate comprehensible ecological thought that will invite dialogue between different communities and cultures. A productive dialogue with diverse audiences on alternative circuits of visual production depends on the base communities' participation and not on the television set or the commercial circuits for renting/selling DVDs. Good Life in this context is not a concept: it is a way of life.

Another challenge that will help to connect personal media projects with a more expansive framework of media production and ciruclation is to find an efficient method to select and make available quality ecofilms. That kind of circulation depends on community groups that take charge of the screening and discussion sessions instead of mainstream production companies. Due to the possibilities from such forms of alternative access, I would like to stress the need of elaborating a basic list of 'ecofilms' that at a minimum show the possibilities of cultural plurality in Good Living. This open aspect of counter-cultural media may enable it to develop into a good archive of models for sustainable communities.

Ecocinema and Good Life in Latin America

Latin American ecocinema reinforces the vital concept of Good Life with all kinds of images and narratives about human and non-human beings. This cultural production goes against the grain and unmasks the shameful lies that economic powers slip into their convincing incentives and unsustainable requirements that ultimately turn to the exploitation of Earth. But we have to be careful with this notion of Good Life in the particular context of ecocinema. It must not be seen as a Latin American variety of 'human development', or another well intentioned, modernized way to interpret attempts to provide a voice for the exploited peoples of the Andes. As Gudynas observed, Good Life cannot be 'ingested' and co-opted by conventional views (2011). The key notion of place in the ecofilms commented on could be understood as evincing a limited environmental awareness. It is also necessary to cultivate a sense of planetary interconnectedness discussed by Heise, 'a sense of how political, economic, technological, social, cultural, and ecological networks shape daily routines' (2008: 55). Regardless of its place centricity, the notion of Good Life will bring new elements to the repertoire of intellectual resources available to the ecocosmopolitan approach that Garrard sees as a third important challenging paradigm-shifter of contemporary ecocriticism (2010: 26–28).

Ecocinema gives a human face to an ecocentric value system that challenges dominant ideologies, which cherish 'development' and insist on eliminating ecology from the political arena. More powerful than the discourse of the practical rationality of 'progress' is the honest portrayal of the people who organize to defend natural resources, recognizing these resources as a vital part of their identity and survival. Ecocinema portrays these people as builders of an alternative way of life in which the key active component of the community is their egalitarian ethics and social philosophy. In their struggles for environmental justice they articulate a dialogue with key participants, such as governments, non-profit organizations, dissident corporate members, and film-makers, among other potential allies. In this context, experimental videos about nature highlight those valuable functions of nature that society ignores, such as purification of air and water, cycling and moving of nutrients, generation and preservation of soils and renewal of soil fertility, and other forms of maintenance of biodiversity like seed dispersal and pollination of crops (Kricher 2009: 186).

Ultimately, the responsibility lies with the spectators, who should practice a new way of listening, a way that considers the agency of non-human animate and inanimate beings and objects, whose voices the dominant rationality has traditionally denied. Some ecofilms show glimpses of the sustainable life we need and they deal with the conflict that originates from the globalized and corporate world. Paying attention to what people can do against the destruction of natural habitats is one compelling way to undermine the common habit of ignoring such crucial ecological conflicts. Another positive approach is to sustain the habit of listening to natural agents that interact with indigenous communities such as mountains, rivers, and birds. These ecofilms not only provoke thoughts about ecological and political actions, they also move audiences to overcome the insensitive way in which society ignores nature and treat any natural resource as an externality.

References

Alvarez Zambelli, C. (2009), *La tierra es nuestra/The Land is Ours*, Valencia, Spain: ACSUR-Las Segovias.

Arana Zegarra, M. (2008), 'Los recursos naturales como mercancía'/Natural Resources as Merchandise, In: *Territorios y recursos naturales: el saqueo versus el buen vivir/Territories and Natural Resources:Plunder versus Good Life* by B. Delen, Quito: Agencia Latinoamericana de Información (ALAI), pp. 11–15.

Bernstein, D. and A. Slick, (2007), *Después de la neblina/When Clouds Clear*, New York: Clear films.

Boyd, S. (2010), *Cuando la tierra llora: Operación diablo/When the Land Cries: Operation Devil*, Development and Peace/Asociación Guarango.

Córdova, A. and Salazar, J. F. (2008), 'Imperfect media and the poetics of indigenous video in Latin America', P. Wilson and M. Stewart (eds), *Global Indigenous Media. Cultures, Poetics, and Politics*, Durham and London: Duke University Press, pp. 39–57.

De la Cadena, M. (2010), 'Indigenous cosmopolitics in the Andes: conceptual reflections beyond "politics"', *Cultural Anthropology*, 25: 2, pp. 334–70.

Fatheuer, T. (2011), 'Buen Vivir. A brief introduction to Latin America's new concepts for the Good Life and The Rights of Nature', Berlin: Heinrich Böll Stiftung/Publications Series on Ecology 17, Translator: John Hayduska, http://www.boell.de/publications/latin-america-buen-vivir-12636.html. Accessed January 2, 2012.

Fernández, P. and Henríquez, A. (2008), *La voz mapuche/The Voice of The Mapuche*, http://www.cultureunplugged.com/documentary/watch-online/festival/play/6068/The-Voice-of-the-Mapuche. Accessed September 30, 2011.

Forns-Broggi, R. (forthcoming), *Handbook of Ecological Imagination of Our Americas*.

—— 'Jungle Tattoo', http://www.youtube.com/watch?v=n5sCf14_ZUM. Accessed July 10, 2009.

—— (2006), 'The House of Wisdom', *After the War: Life Post-Yugoslavia*, A Million Movies a Minute.

—— (2011), 'The bird's dream', In ASLE (Association for Studies of Literature and Environment), Ninth Biennial Conference, 2011 'Species, Space and the Imagination of the Global', 21–26 June, University of Indiana: Bloomington.

Garrard, G. (2010), 'Ecocriticism', *The Year's Work in Critical and Cultural Theory*, 18, pp. 1–35. Downloaded from ywcct.oxfordjournals.org at Aurary Library. Accessed November 3, 2010.

Gudynas, E. (2011), 'Good Life: germinating alternatives to development', *America Latina en Movimiento* 462, Traslator: Bob Thompson, http://alainet.org/active/48054. Accessed December 30, 2011.

Gudynas, E. and Acosta, A. (2011), '*El buen vivir más allá del desarrollo*/Good Living Beyond Development, *Quehacer* (Lima, Perú) 181, pp. 70–81.

Havránek, V. (2008), 'The documentary: ontology of forms in transforming countries', M. Lind and H. Steyerl (eds), *The Green Room: Reconsidering the Documentary and Contemporary Art*, Berlin and New York: Stenberg Press/Center for Curatorial Studies at Bard College (CCS Bard), pp. 128–43.

Heise, U. (2008), *Sense of Place and Sense of Planet: The Environmental Imagination of the Global*, New York-Oxford: Oxford University Press.

Ivakhiv, A. (2008), 'Green film criticism and its futures', *Interdisciplinary Studies in Literature and Environment*, 15.2, pp. 1–28.

—— (2011), 'The anthrobiogeomorphic machine: stalking the zone of cinema', *Film-Philosophy*, 15.1, pp. 118–39.

Li, F. (2009), 'When pollution comes to matter: science and politics in transnational mining', Ph.D. thesis, Davis: University of California.

Lu, S. H. and Mi, J. (eds) (2009), *Chinese Ecocinema in the Age of Environmental Challenge*, Hong Kong: Hong Kong University Press.

Kricher, J. (2009), *The Balance of Nature. Ecology's Enduring Myth*, Princeton and Oxford: Princeton University Press.

McClennen, S. (2008), 'The theory and practice of the Peruvian Grupo Chaski,' *Jump Cut. A Review of Contemporary Media*, http://www.jumpcut.org/archive/jc50.2008.Chaski/index.html. Accessed February 15, 2012.

Monani, S. (2011) 'ASLE 2011 Ecomedia Seminar', Ecomedia Studies seminar, ASLE, Bloomington, June 2011, http://asle-seminar.ecomediastudies.org/. Accessed July 26, 2011.

Olmos Morales, L. (2008), *El llanto de la tierra/The Cry of the Earth*, http://www.youtube.com/watch?v=PUS69Ev6HwE. Accessed October 14, 2011.

Overing, J. and Passes, A. (eds) (2000), *The Anthropology of Love and Anger: The Aesthetics of Conviviality in Native Amazonia*, London and New York: Routledge/Taylor & Francis Group.

Salgado, F. (2010), 'Sumaq kawsay: the birth of a notion', *Cadernos EBAPE.BR* 8.2, pp. 198–208, http://dx.dol.org/10.1590/S1679-39512010000200002. Accessed December 30, 2011.

Schiwy, F. (2009), *Indianizing Film: Decolonization, The Andes and The Question of Technology*, New Brunswick, NJ and London: Rutgers University Press.

Smaill, B. (2010), *The Documentary: Politics, Emotion, Culture*, London: Palgrave MacMillan.

Solanas, F. (2009), *Tierra sublevada: Parte 1, Oro impuro/Land in Revolt: Part One, Impure Gold*, http://www.youtube.com/watch?v=Cl8wmDizLWo. Accessed September 27, 2011.

Tótoro, D. (2004), *Üxüf Xipay-El Despojo/The Plunder*, http://video.google.com/videoplay?docid=1337356709020319057. Accessed October 12, 2010.

Watts, M. (2009), *Laguna negra/Black Lagoon*, http://www.cultureunplugged.com/documentary/watch-online/festival/play/6225/Laguna-Negra. Accessed September 14, 2011.

Dimensions of Humanity in *Earthlings* (2005) and *Encounters at the End of the World* (2007)

Ilda Teresa de Castro

The division or *scissura* that marks the relationship between the contemporary human being and the cosmobiological dimension of life on Earth, and the alienation that this division has created, results in the near impossibility of changing the circumstances that lead to the progressive depletion of natural and environmental resources. The difficulty imposed by this *scissura* grants in conventional and equally distant regulations a management of living that—in the mechanics of convenience economics from which it has emerged and the attitudes of political correctness that it brings—embodies the whole ineffective facade of good intentions that the worsening of the global climate evidences. Thus, the need for a new ontological primacy reliant on new theories of existence is increasingly evident in evoking perspectives that could lead to a reformulation of established models and to new paradigms. We can recognize the emergence of thoughts and theories that with a certain sense of urgency reflect on the state of humanity and civilization. Inseparable from the idea of the collapse of belief in endless progress, advancement, and technological development *per se*, they register the loss of seminal values, and seek to resituate the current model of consumption, hyperindustrialism, globalization, and massification of society. They aim to recuperate the preservation of an organic and existential spiritual dimension of the human species. That is how, in the dawn of the second decade of the twenty-first century, during which everything is vulnerable to the risk of extinction of specie(s) and civilization, there persists in the human being the urgency of the question: Can there be another way of life and how can we go about creating this alternative? In this reformulation, still without definite shape, the revaluation of individual and collective consciousness is fundamental for fostering a *new look* at the problem of existence, and to creating an *art of living* that promotes new existential values. Crucial to this is the creation of new methodologies to perceive the world in ways that consider the existence of the subject in the injunction of body and soul, and in its multiple values.

The role and potential of cinema in framing planetary realities in a way that engages and encourages awareness and cosmobiological ecocriticism on the part of the viewer is particularly strong at a time when the survival of several species, including humankind, appears particularly vulnerable to the global climate (and civilization) crisis. These ecocinematic concerns lead to the comingling of distinct cinematographic lineages. The most obvious is the interaction between mainstream US-based ecocinema and films from different backgrounds, which do not share the same foundation and structure of production and dissemination. Illustrative examples of this can be found in *An Inconvenient Truth* (Davis Guggenheim, 2006) and *The Day After Tomorrow* (Roland Emmerich, 2004), which contrast with, for example, *Meat the Truth* (Gertjan Zwanikken, 2008), or *Spring, Summer, Fall, Winter ... and Spring* (Kim Ki-duk, 2003). Another possible distinction exists between ecoactivist films and films that through their creative and philosophical idiosyncrasies reveal, reflect and embrace the subjectivity of the director's personal universe—such as is the case with *Nanook of the North* (Robert Flaherty, 1922), *Au Hasard Baltazar* (Robert Bresson, 1966) and *Le Rayon Vert* (Eric Rohmer, 1986). When categorizing ecoactivist films I include films that incorporate ecocritical information as a unique leitmotif, and which are generally more representative in their content rather than due to the formal cinematic techniques that they employ. *Earthlings* (Shaun Monson, 2005), *An Inconvenient Truth*, *The 11th Hour* (Leila Conners Petersen/Nadia Conners, 2007), *Meat the Truth* and *Home* (Arthus-Bertrand, 2009) are some examples of films that fit this category. *Home* is especially useful in illuminating these concerns as its main subject matter concerns the Earth's natural elements and its actors are those elements with whom humanity interacts. The focus is on life, its origin and the transformations that human activities have instilled in certain moments of this natural history, one which parades itself on screen as an art history of nature. Arthus-Bertrand states in an interview[1] that his decision to direct this documentary was made as he was introducing a screening of *An Inconvenient Truth* in the French Parliament. He understood the impact that cinematographic works can have on an audience larger than the one obtained with TV shows—and how much a feature documentary can 'reach' people.

In any case, the key issue involves cosmobiological awareness, the path of change—to affect solutions on a local and global scale—and in this procedure, the reception and dissemination of ecodocumentaries that approach both local and planetary ecological realities in their comprehensive contamination are crucial.[2] In this sense, Murray and Heumann (2009) note a greater prevalence of films with political and environmental content starting in the 1990s, accompanied by interest in publications such as *The Hollywood Reporter* and *Variety*, and which, according to the analysis of *The UCLA Environment Report*, have helped increase the number of stories focusing on environmental issues. In addition, the Environmental Media Association (EMA) Awards have included since 2004, 'a separate category for environmental "process" improvements based on EMA's Green Seal checklist' alongside categories related to films and TV shows with environmental content (Murray and Heumann 2009: 3). If the assessment of that impact confirms the pertinence of cinematic ecologism, the relevance of ecocinema cannot be separated from the environmental footprint of the film industry. As Nadia Bozak recalls, the environmental impact of digital productions is much less than the one of a feature documentary using film stock, and it can be configured into a self-sufficient and ecological mode of expression and communication (2012). Simultaneously, as emphasized by the conclusions of the extensive document 'Sustainability in the Motion Picture Industry', produced by UCLA in 2006, the complex set of environmental relations behind any image in motion need to be approached critically, a notion from which 'carbon emission-free films' cannot be detached, and of which *An Inconvenient Truth* or *Home* are examples. Even though the focal point of Murray and Heumann's book, *Ecology and Popular Film*, is mainly on mainstream American production, the authors conclude that, if Hollywood adopts green production methodologies, new perspectives will surely arise in ecocritical film studies, which will also increase the possibility to influence readers and viewers towards ecoaction.

In addition to Hollywood perspectives, it is not difficult to note a range of diverse agents and concerns in global film culture. As an example, the plot of the documentary film *Climate Refugees* (Michael P Nash, 2009), selected for 24 film festivals in 2010, including Sundance and Cannes, revolves around the problem of environmental refugees, 'the human face of climatic changes'. Evidencing the transversal and transnational scope of ecological concerns, this is a predicament at the basis of the 2008 exhibition *Terre Natale, Ailleurs commence ici*, where both Paul Virilio and Raymond Depardon participated. The urgency and relevance of these perspectives is something which the ecoactivist films share, as they call for raising awareness as a precursor to transformative actions, and laying ground for the acceptance of potential drastic transformations in the near future. *The 11th Hour* is a good example of a cinematic contribution to this mission. The fact that the filmic structure relies on first-hand statements of specialists, scientists and authors around whom the narrative unfolds, provides it with a semblance of credibility. While *An Inconvenient Truth* revolves around Al Gore's character, his life and environmental concerns, and both its factual information and ecocritical perspectives are inserted into that context, in *The 11th Hour* the structure is regulated by research and scientific criteria. The film weaves an informational mesh of substantial scope out of the face-to-face interaction of economists, ecologists, scientists, designers, writers and politicians. The scientific expertise of everyone involved is notable and frequently coincides with view-points about new scientific approaches to life and advanced theories by authors such as Fritjof Capra, Buckminster Fuller and Gregory Bateson. Thus, in addition to the assessment and information about actual ecological imbalances and their possible causes, something common to both films, *The 11th Hour* activates reflection and provides information about critical issues which are insufficiently clarified not only in *An Inconvenient Truth* but also in wider cultural ecopolitics. What are the main reasons for the non-implementation of solutions? Which forces are blocking change or which solutions are proposed?

Meat the Truth is another example of a documentary film with an ecocritical mission, an emphasis focused on the animal question. Even though it is not a case of docufiction, since it is always faithful to the documentary genre, it still incorporates mechanisms of fiction in its narrative construction, and gradually reveals its fictional nature, creating from the beginning a sense of 'something concealed' that in time becomes an implied *leitmotif*, just as important to the film as the collecting and displaying of information on its main theme. Performing as themselves is fundamental for Al Gore as it is to Marianne Thieme in *Meat the Truth*. She is a member of the Dutch Parliament representing the Animal Party, in a country that was the first one in the world to defend non-human animal rights in a National Parliament, and it is her personal involvement with the thematic issues at stake that are crucial for the credible staging of the film—much as is the case with Gore's presence. Curiously enough, the staging of *Meat the Truth* replicates the staging of *An Inconvenient Truth* in several respects: the guiding thread of the presenter on stage in front of an audience, the articulation complemented by archival video footage, the introduction of an animated replica of *The Matrix*, the *Meatrix*, similar to the presence of the feminist cartoon character Lisa Simpson in Gore's film, Thieme's gag where she climbs a graphic in the audience room so as to mimic Gore. Furthermore, the film's subtitle, *A Truth More Than Inconvenient*, directly references the former, the meaning of which is revealed as the narrative unfolds: *Meat the Truth* introduces one of the largest contributors to global warming that is not mentioned in Gore's film: livestock farming generates more greenhouse gas emissions worldwide than all cars, lorries, trains, boats and planes added together and, according to the producers, *An Inconvenient Truth* ignores this fact due to Gore's personal involvement in livestock production.

In contrast to these personalized econarratives, the inclusion of actor Leonardo Dicaprio, who is also one of the producers, as presenter in *The 11th Hour* is discrete, sometimes describing the film's intentions, sometimes combining the advanced conclusions of field experts with the introduction to the next segment. Still, he does not control the development of the narrative as happens in *An Inconvenient Truth* and *Meat the Truth*, nor does he fulfil the role of the guiding thread performed by the voice-over in *Earthlings*—or even *Encounters at the End of the World*. In *The 11th Hour*, the presenter is in fact an actor who assumes his real identity when performing the role of the presenter. Likewise, in *An Inconvenient Truth* and *Meat the Truth*, the presenters, non-actors, also assume their real identity while performing their role as presenters of the film, whereas simultaneously, as main characters of the film, and also due to their heavy media presence, they elevate themselves from their role as non-partisan non-actors.

The scope of ecodocumentaries is substantial and can evince a completely different structure to depicting ecosystemic concerns. Ecocritical denunciation sometimes finds echo in films that, even though they do not have that mission as a goal, end up to fulfilling an influential role in global ecological film culture. That is the reason this article analyses the filmic content of two completely different features, one arguably an ecoactivist film, *Earthlings*, and one that cannot be put into that category, *Encounters at the End of the World*. They are both cases which emphasize their reduced material footprint, as they are digitally produced, use natural settings, are shot with handheld cameras, feature non-actors in low-budget productions and produced with a reduced film crew. In *Encounters*, the crew consisted of a sole cameraman, besides Herzog, who was also in charge of sound capture. In *Earthlings*, parts of the recorded images were shot by Monson, frequently with a hidden camera, while the other part consists of archival footage from other independent productions.

Both are also recent works that present an intimate glimpse and specific mapping of reality, establishing a filmic order directly related to Earth and the living, and through their ways of seeing instil new ways of seeing in the viewer. These works represent *scissuras* from established paradigmatic and conventional structures. Both films show how the implications of environmental issues and living in the precariousness of the present situation have spread beyond geopolitical and economic boundaries, and exhort human beings, states and nations to global and cosmobiological awareness. In the movement that they both project, they envision a space that can make room for a global perspective that positions itself inside and outside of the reality that it observes, while it is also a part of that very same reality. In the case of *Earthlings*, this global perspective is also present in the reflection and analysis of the use of non-human animals for the benefit of humans, while in the case of Herzog's film, it emerges as a part of the community that lives away from consumer societies. These are different looks at economic interests, methodologies and practices of global capitalism, questioning it through the way they display it and the way they make contact with it. And this movement is the basis of a consciousness that is simultaneously local and global, individual and collective.

Earthlings

Earthlings may be considered one of the most realistic horror films in the history of cinema, amounting to what is something akin to 'an anatomy of horror'. The film exposes the greatest plague unleashed on the living in the late modern era and one of the most horrendous acts in human history, resulting in the greatest number of victims. The film, which focuses on the plight of exploited animals in the food and clothing industries, domestication, entertainment and scientific research, was shot in kennels, pet stores, industrial production units, abattoirs, leather and fur trade areas, animal shows and sporting activities, circuses and research laboratories. It bears witness to the daily barbaric handling, massacre and torture of billions of non-human animals, acts promoted and executed by humans and justified as being for the benefit of human subsistence, incorporated as part of the consumption of other animals. Some are aware of the facts about these methods, others do not know of them, and still others prefer to ignore them. This is also aggravated by the fact that these events are coexistent with the present, a continuing and condoned process, although sometimes without the full knowledge or awareness and consent of individuals who for the most part do not participate directly in the processes that are the basis of production. This is why the film states that if humans had to kill to eat, almost everyone would become vegetarian.

The film begins with a written message on a black background: 'The images you are about to see are not isolated cases. These are the Industry Standard for animals bred as Pets, Food, Clothing, for Entertainment and Research. Viewer discretion is advised'. This is a departure from the more common 'This is a film of fiction, any resemblance to reality is purely coincidental'. In contrast, *Earthlings* advises that 'these are not rare and isolated cases, this is real and this is the norm!' Complementing this assertion, a second written message on a black background appears, combined with the introduction of Moby's calm music that will accompany the entire film:

The three stages of truth

1. Ridicule
2. Violent opposition
3. Acceptance

By referring the viewer to the classification of truth in three stages, the film engages immediately with the field of philosophy, ethics, the question of what the 'truth' is and the relationship with which it is intertwined—these are also the early stages that anticipate the most common reactions of the spectator confronted with the content and the proposals that the film sets out. The first actual footage are close-ups of Earth interspersed with the film's title, images that, through the visual beauty they document, and the soft, calm music accompanying it, inspire the viewer to a communal identification with the planet, also bringing to mind the 'Spaceship Earth' of Buckminster Fuller, an organic spaceship that transports everything and everyone who inhabits and lives in it. There are successive close-ups of the planet, and since the film is about Earth and the behaviour of all its inhabitants, characters are defined in the following shot of the title-word 'earthling, 'earth.ling *n*. One who inhabits planet Earth': the film is about them, inhabitants of planet Earth, earthlings, human and non-human.

The opening sequence of the film acts as a starting point and a point of reflection for the whole work. Throughout the film, the quiet tone established by Moby's music and the halting voice-over narration by Joaquin Phoenix forms a support base for an alternating assembly of long and short cuts, archival footage and recordings that establishes a homogeneous background into which Monson inserts shocking testimonies of cruelty. An example of this is the one-shot scene following the running of the bulls in Spain as we see a bull tied to a string by the horns, which tries in vain to free itself. With its movements limited the bull drags its horns through the straw on the ground and comes up against the fence which separates the bull from the audience. We hear thrilled and enthusiastic cries from the spectators. A group of men surrounds the bull, violently pulling and pushing while the bull struggles. The shot fades to black and stays black with the sound of voices from the wild crowd.

Through this spectacle of cruelty, the film parallels racism, sexism and speciesism, which the historical images of slavery, the Ku Klux Klan, Nazi rallies and suffragette demonstrations establish, which alternate with contemporary shots that highlight a second example of abuse of non-human animals: the imprisonment of caged and stacked piglets. The shots of archive images that follow document cases of human-on-human violence, and prepare the viewer for the third example of the abuse of power by humans over non-humans: the crushing of the skull of a pig immobilized in the soil. A man approaches a pig lying on the ground, grabs a cement brick and throws it onto the head of the pig. From the recording quality of this sequence we deduce that the image was captured with a hidden camera.

The opening sequence, which articulates parallels between human and non-human earthlings, and also emphasizes the opposition between imprisonment and freedom, well-being and suffering, summarizes the modus operandi and intentions of the film. Switching between peaceful and shocking scenes, Monson always establishes a lilting, diegetic rhythm of cadenced homogeneity. The merging of shocking and tranquil footage allows the film to establish an emotional balance in the viewer which fulfils the need of keeping the filmic within certain limits of supportability. That is, the important content, which is shocking *per se*, avoids an overload of shock by means of cinematographic resources. This caution is also present in the warning in the opening shot: 'Viewer discretion is advised'. Thus, this alternation between scenes that are peaceful, representative and evocative of harmonious worlds, structures only the initial and final sequences—it is not a feature that is maintained throughout the film.

The five thematic chapters that follow—pets, food, clothing, entertainment, medical, scientific experiments—show the different ways in which non-human animals are used by humankind. All of these chapters denounce behaviour that is aberrant and absolutely inhuman in its abuse, which in turn invokes the moral and ethical responsibility of all humans for these practices perpetrated on non-human animals, whether by academics or by butchers. The recorded situations are accompanied by a descriptive and informative voice-over in a calm, peaceful tone of voice, which provides statistical data to complement the images with which the audience are being presented. Always maintaining an informative character, avoiding any kind of emotional involvement, it merely seeks to clarify what the images show.[3]

The assembled records show the practices exercised by humans over non-human animals in the framework of anthropocentric thought and the unsustainability of consumption policies, economics, and industrialism introduced in recent centuries in developed societies. In addition to establishing a line of ethical questioning, Monson also reports on the ecological impact of the procedures that have proved to be deeply damaging on a planetary scale. He affirms how one of the main causes of pollution and environmental degradation—the intensive industrial production of meat—includes all consumers on a global scale and depends on participation that is both individual and global and clearly transnational. Underlying all the approaches to violence and exploitation of non-human animals that Monson stresses is awareness of the commonality that unites all beings that exist and live on planet Earth, regardless of any trait that distinguishes them—i.e. being an earthling, an inhabitant of Earth—which precedes other classifications constructed by humanity.

The difficulty of accepting the practices that the film uncovers and gradually instils in the viewer is increased by simultaneous confirmation that these situations that assault us may not be difficult or unbearable for others, which is no doubt true of those who perpetrate the acts revealed on the screen. In this way it poses immediately the question of insensitivity and cruelty manifested by those involved—authors of different types of cruelty performed on these non-human animals and about standards that enable and allow these behaviours. Permissiveness exists in the chain structure: not only are the individual and group behaviours enabled and allowed—even in concealed or secret legality—but the very processes of production in which they operate are perpetuated over and over and accepted, legitimized by the mechanics of production and consumption. In this way, the characteristics of the anthropocentric, mechanistic and capitalist model of society developed in modernity and contemporaneity appear to be interstitially attached to these procedures in space-time, granted consent through the total disregard for what is not human and the absolute primacy of profit, money and power over the other.

The movement of awareness that these records cause in the viewer is dependent on the shock perpetrated by the perception of permanent coexistence of these practices in real life and the involvement of implicit consent in these routines. Yet, the relationship that the viewer shares with the issue of implied *consent* is not completely clear. Wrapped up by the rules of consumption in assisting the perpetuation of these procedures, the viewer is positioned to debate the idea of guilt arising from the confrontation with these realities. Therefore, one of the reactions is the frequent refusal to accept involvement: claiming ignorance of them or invoking the objection that these practices are not present in all geographic areas. Alternatively, this takes the form of seeking to escape the scourge of one's involvement by denying the homogenization of suffering in the practice of animal exploitation in the global and transnational consumer society in which one participates. However, viewers can not escape the fact that the current supply market is global and open, promoting free movement and marketing of products from all geographic areas, nor the fact that the reported cases arise primarily in the United States, but also in China, Japan, Canada, and Europe. But simultaneously from an emotional point of view, the viewer avoids confronting and opening the imagination to what Cora Diamond (2008), referring to the drama of Elizabeth Costello, identifies as 'living' in the body of another, i.e. the ability to feel another's pain, to put oneself in the place of another living body. In the case of Costello, the renowned writer feels haunted by the horror of what is practiced by humans over other animals to feed themselves and she becomes wounded by this knowledge of the complete insensitivity of other humans in this matter. With a troubled mind and jangled nerves, she is 'living' in her own body as the body of a wounded animal, 'not a philosopher of mind but an animal exhibiting, yet not exhibiting, to a gathering of scholars, a wound, which I cover up under my clothes but touch on in every word I speak'[4] (Cavell 2008: 107).

In case of the vision of *Earthlings*, the awareness that the film raises of the realities unthinkable for most viewers builds a space of exposure that is hard to leave without some affective impact. The atmosphere of pain and entrapment infects the viewer who is confronted with the representation of the crimes in which he/she is complicit, and through it, refers him or her to the problem of connecting perception with consciousness. The film asks us to feel the presence of fear, pain and distress in the animals seen, since

[T]he animals may have a particular perception of the atmosphere: when they 'feel' the approaching death, they show a keen sense beyond the common perception that in most cases is beyond human beings. 'Feeling' death as 'feeling' fear is the same as understanding forces that associates the animal as a determined phenomenon.

(Gil 2005: 18)

This unbearability, which can lead the gaze or break the projection or viewing, activates a will to alternative choices to escape that pain and suffering but also the awareness and action arising from the intimate acknowledgment of the subject, since not acting equates to a kind of complicit consent. The extremes of brutality that *Earthlings* documents are exemplified in the operation of skinning animals alive—so that the integrity and quality of their skin is kept—after they are kicked and trampled on the ground, or beating to death a group of dolphins captured using the knowledge that when one dolphin is hurt the rest will not abandon it.

Regarding the *pathos* implicit in this filmic content and in this reflection, I recall the words of Derrida: 'If these images are "pathetic", if they invoke sympathy, it is also because they "pathetically" open the immense question of pathos and the pathological, precisely, that is, of suffering, pity and compassion' (2006: 47), and without forgetting that the emergence of this state in the viewer comes from the compassionate ability to put oneself in the shoes of another, to feel beyond our own body and beyond the human body. But even if the viewer does not adopt this compass of affection, the depth of the cruelty involved in the acts he/she witnesses on the screen, in itself, highlights the great *scissura* reached between the ability (and practice) of the domain of the weaker and the consciousness (and practice) of the ethics of the living. The profound ecological and human impact this separation of values invokes reflection on the interrelationship between the serious ecological imbalances and the current imbalance in contemporary human evolution. In fact, the mechanisms of hyperindustrialized consumer society have allowed the dehumanization of the human articulated by market convenience and the alienation of the values of living in their most essential and ancestral features. The hegemony of business has taken over the direction of an alienating logic of life that is hard to escape (Weber 1967) and of which we are hardly aware, i.e., the de-individualization promoted by the dominant political and economic systems culminates in the states of often unconscious alienation where the human is dehumanizing himself/herself. This panorama has created the emergence of the need for an ethical individual and collective development and proportion between progress and spirit, mind, and the self—as Guattari (1989) and Stiegler (2004) explore in their works. At the present stage, there is need for the (re)creation of another arena for increasing human awareness. Films like *Earthlings*, which call for individual introspection and collective analysis and accountability that is both local and global, can provide information that is otherwise difficult to access.[4]

In this specific case, and unlike other ecoactivist films, Monson does not supply the viewer with a list of solutions or behavioural rules to be adopted like *The 11th Hour* does. Here, the obvious suggestions emerge from the filmic content itself and consist of practices involving non-human animal suffering; i.e. it is through the witnessing of these descriptive feature stories that the viewer becomes acquainted with the realities from which they are usually set apart or unaware. It is through the empathetic embodiment these revelations instigate that the feeling of shock is activated and the process of awareness and accountability occurs. Although *Earthlings* is a shocking film, it is so precisely because it shows what is usually hidden, a type of information whose disclosure is not convenient for the powerful established economic interests it puts at stake.

The action in *Earthlings*, as well as in so many other documentary films of shocking denunciation, has had positive outcomes. For example, in Great Britain, Animal Aid has produced the documentary film *UK Slaughterhouses* (2009) on the conditions and methods used in English slaughterhouses. Its images, secretly shot, revealed abject realities and practices, and have been used as evidence in denunciatory campaigns.[6] In spite of the shock it may cause, *Earthlings* is based on real images, complemented by a thorough collection of real statistical data, something that substantiates its legitimate character, relevance, and recognized urgency, namely as a pedagogical film for young and adult audiences. It is a film that makes the interdependence between environmental issues and human behaviour self-evident, as well as the manipulation and the withholding of information undertaken by capitalist consumer societies. The play on proximity and distance that the film constructs aims to activate reflection and questioning by viewers in respect to the world of which they are a part and in which they participate, instigating awareness and a change of attitude.

Encounters at the End of the World

In a completely different context, *Encounters at the End of the World* calls for another type of ecoawareness in the proximity/distance connection it establishes between its narrative structure and the viewer. The film focuses on a set of individuals in a unique area of the world, Antarctica, or the 'end of the world', as Herzog puts it with some degree of affection and irony. By circumstances specific to each of them, they are particularly sensitive to new values of civilization and a new orientation in the relationship with the non-human which they encounter during their research. It is through the protagonists that the film promotes observation and reflection. For instance, the glaciologist Douglas MacAyeal explains how scientists have acquired the ability to recognize ice as a living and dynamic entity, such as the B-15 iceberg that he studies. The cutting-edge scientific experiments he is involved in have led him to assert that the traditional view of Antarctica is obsolete: the current view is that of a living being that produces dynamic changes that have repercussions for the rest of the world, perhaps in response to the effect that the world has on Antarctica.

Throughout the film, Herzog connects voice-over commentary with interviews with those who live at McMurdo Station and its surrounding area. Through a constant critical perspective and irony that have become characteristic of his documentary films, the narrative serves as a guideline to the description of everything he comes across during his journey. Herzog also permeates this with ecological questions, as he wonders what will happen in the future after the disappearance of the human species and whether there will be archaeologists from other planets trying to understand what humans were doing at the South Pole. If they go down the tunnels dug in the ice where the temperature is around 70° below zero, they may find a huge frozen sturgeon or a picture of a world once green, memories of human presence on the planet.

The individuals the film follows espouse different perspectives on reality based on their specific areas of expertise, and because they are living at a considerable distance from the alienating mechanisms of today's dominant models. In an absolutely relevant way, their perspectives enable the viewer to imagine himself/herself as living away from urban and mainstream mechanisms of civilization, and still being a part of an educated society. The community of scientists develop cutting-edge scientific research correlated with life in its most elementary states—a place where some of the most advanced scientific research comes together. As the film itself is a journey, it depicts subjects who are themselves on the road, contemporary nomads; they cross borders until they reach one of the most distant points of the planet and there, separated from their national and cultural roots, they share their interculturality. This reality is embodied in the words of Stefan Pashov, a philosopher who works as a forklift operator, as he defines himself as an explorer of mental landscapes and worlds of ideas, and justifies the presence of these unique characters at the 'end of the world' as the 'natural selection of people who have the intention to jump to the edge of the map' and are at the point where all lines converge, since there is no point south of the South Pole. These are the 'professional dreamers, those who dream all the time and through which the cosmic dream is realized', he states. Or the linguist William Jirsa, who, disillusioned with 'the stupidity of academia' and its indifference to the risk of the disappearance of minor languages and the cultures they represent, has built a vegetable greenhouse and has found a group of people he recognizes as a family of elective identity. 'PhDs who wash dishes, in a continent with no language', he reports with a smile.

Taking into account the remarkable work Herzog does with the landscape—as well as with irony—Eric Ames (2009) notes the vital importance of landscape for Herzog's narrative features and documentary films, where the correlation between outside and inner landscapes is used as a recurrent methodology through which Herzog claims his connection to Caspar David Friedrich, namely when he states that:

> For me, a true landscape is not just a representation of a desert or a forest. It shows an inner state of mind, literally inner landscapes, and it is the human soul that is visible through the landscapes presented in my films, be it the jungle in *Aguirre*, the desert in *Fata Morgana*, or the burning oil fields of Kuwait in *Lessons of Darkness*. This is my real connection to Caspar David Friedrich, a man who never wanted to paint landscapes *per se*, but wanted to explore and show inner landscapes.
>
> (Ames 2009: 51)[7]

In *Encounters,* the white and blue Antarctic landscape is also very singular, and the researchers Herzog finds in it are biologists, volcanologists, physicists, and explorers, people who face extreme conditions of survival, who plunge into the waters of glaciers where they seem like astronauts floating in space, who climb into funnels of ice in the heart of Earth where the white cathedrals are waiting—a poetic description that corresponds to the images the film displays. The film-maker points out the analogies between the routine with which the divers prepare to dive in silence and the preparation of priests for the homily, between the reality beneath the ice where space and time acquire a strange new dimension and the ambience of cathedrals (which recalls the emphasis on the sublime and sacred he explores in his work). This leads him to ironically wonder if extraterrestrial explorers will still find some ice or if they will have to build an artificial Antarctica in a studio, recalling that modernity has preferred the production of artificial ice, and to this purpose invented and produced appropriate machinery.

Following this philosophical line of inquiry, Herzog comments on the parallels between the pathway of cosmologists in search of the origin of the universe and of scientists tracing the evolution of life to its points of origin. Examples of this are various projects undertaken by scientists who share his interest in new perspectives and scientific approaches to the living. Examples involve the research on *foraminifera* trees, primordial unicellular microorganisms that exhibit such behavioural qualities that they earned the right to be proposed as a primary form of intelligence at the end of the twentieth century, when the British scientist Heron-Allen acknowledged the possibility of applying to them all formulated definitions of intelligence. Or the investigation and detection of the most undetectable subatomic particles, known as *neutrinos*, which are able to traverse matter without any interference, as if they existed in a separate universe. Neutrinos were the dominant particle a few seconds before the Big Bang and determined most of the connections in the production of the elements. The scientists, who try to establish contact with this other world, the world of neutrinos, describe them as spirits because they cannot be touched but can be measured: 'it's like measuring the spirit world!' Peter Gorham, a physicist responsible for the project says.

But one of the most surprising situations—obviously for a human way of seeing—that *Encounters* allows us to witness is the decision made by penguins to commit suicide, as they move away from the rest of the community and go alone to the distant mountains towards certain death, challenging human assumptions and understanding of other animals. David Ainley, a marine ecologist who has studied them, explains that even if they are interrupted in their course and put back in their colony, these penguins immediately resume the previous path of no return. 'But why?' Herzog asks without irony and the viewer also wonders, confronted with the questioning of the anthropocentric models of knowledge that this testimony urges.

Conclusion

Despite the disparity of contexts and other dissimilarities, both *Earthlings* and *Encounters* focus on the conditions of life that different species share on planet Earth, as well as the ethical responsibility that comes with the planetary focus the films espouse. Just like in

Earthlings, despite the presence of non-human animals as the main subject, the focus in *Encounters* is on the human dimension, its influence on the planet and those who live on it. While the film emphasizes the human dimension as its main research goal, what its cinematic content reveals goes far beyond those limits. Herzog seeks to understand the ways of life of those whom he finds living at the end of the world and the profound questions they pose about living and the meaning of being alive. The threat of natural disaster, the failure of the dominant anthropocentric mechanicist paradigm and the capitalist model of the hyperindustrialized consumer, the possibility of returning to (near) essential and original states, the reconsideration of the value of all existing substances on the planet, are some of the themes the film emanates. Such a profound form of questioning was present from the project's inception as Herzog told the National Science Foundation—which invited him to make the film—that he would not come back with a film about penguins, since his questions about nature were different: to understand why humans wear masks or feathers to hide their identity, why they saddle horses and feel compelled to chase the 'bad guy', why certain species of ant use colonies of *psilos* as slaves to produce sugar drops for them, why a sophisticated animal like a chimpanzee chooses not to subjugate inferior creatures, since it could very well mount a goat and gallop into the sunset. Similarly, *Earthlings* intimates recognition of the limits of overcoming the unethical practices that it portrays—curiously, and perhaps by no connivance with economic, political, or social systems, or prejudiced thought, the film was initially rejected by distributors, who warned the director that the film would have no chance of being successful: it would 'never see the light of day and should be swept under the rug'—and the analysis of the interconnection between the exploitation of non-human animals and the economic interests that this exploitation supports stimulates critical, social and political inquiry in the spectator.

These forms of inquiry are inevitably focused on human concerns, but they by no means remain relegated to dominant anthropocentric understandings. Herzog does not hide the very anthropocentric from which he operates when he reflects on the various movements established between the humans, nature and the universe, in local and global realities that encourage a shift of the dominant paradigm. He stresses the specific events in an area of the planet, that are simultaneously both the result of actions carried out everywhere, thousands of miles away, on a global and meta-national scale, and are also themselves potential triggers of new climate and social changes on a world scale. That is, given the ecological changes that are underway, the ecoscape of Antarctica is also symbolically planetary.

The lives that *Earthlings* documents reveal procedures practised in areas, states, nations and diverse cultures, in economic and political occasions on a global and transnational scale. The film proves the existence of connivance between individual and communal, national and transnational processes, overflowing on a global scale that involves institutional and disciplinary practices and policies of the global economy. Monson's independent production process crosses geographic and cultural boundaries and questions a neo-colonial speciesism, which is indiscriminately accepted by many forms of power. This is a process that simultaneously provides an exploration of the state of nations and questions the individual ethics of spectators. Similarly, what best defines the population of McMurdo Station and other parts of the community of scientists that *Encounters* discovers, more than traces of belonging to a nation and its structures, are concepts of civilization, philosophy, ecology, and science. Simultaneously, this group of subjects that moved to this area for professional, philosophical or existential reasons also forms a new population in Antarctica. Coming from different regions of the developed world, they now inhabit McMurdo Station, and have become its local population—and that is how the diasporic experience of McMurdo Station reconfigures the population fabric of Antarctica and establishes a new space of cultural identity and community. Finally, both films converge in evoking interrelationships and conscious connections between vegetable, mineral, and animal nature, between human and non-human, or in the words of philosopher Alan Watts with which *Encounters at the End of the World* ends: '[T]hrough our eyes the universe is preceding itself, through our ears the universe is listening to its cosmic harmonies, and we are the witness through which the universe becomes conscious of its glory, of its magnificence'.

References

Ames, E. (2009), 'Herzog, landscape, and documentary', *Cinema Journal*, 48: 2, Winter (pp. 49–69).

Bateson, G. (2000), *Steps to Ecology of Mind*, Chicago: The University of Chicago Press. Foreword by Mary Catherine Bateson (1972).

Bozak, N. (2012), *The Cinematic Footprint: Lights, Camera, Natural Resources*, New Brunswick, New Jersey, and London: Rutgers University Press.

Cavell, S., et al. (2008). *Philosophy & Animal Life,* New York: Columbia University Press.

Coetzee, J. M. (2004), *Elizabeth Costello*, Lisboa: Dom Quixote.

Derrida, J. (2006), *L´animal que donc je suis*, Paris: Editions Galilée.

Diamond, C. (2008), 'The difficulty of reality and the difficulty of philosophy', in S. Cavell, C. Diamond, J. Mcdowell, I. Hacking and C. Wolfe (eds), *Philosophy & Animal Life*, New York: Columbia University Press, pp. 43–91.

Gil, I. (2005), *A Atmosfera no Cinema—O Caso de A Sombra do Caçador de Charles Laugthon Entre Onirismo e Realismo*, Lisboa: Fundação Calouste Gulbenkian, Fundação para a Ciência e a Tecnologia, Ministério da Ciência e do Ensino Superior.

Guattari, F. (1989), *Les trois écologies*, Paris: Editions Galilée.

Herzog, W. (2007), *Encounters at the End of the World*, Discovery Films.

Monson, S. (2005), *Earthlings*, Nation Earth.

Murray, R. L. and Heumann, J. K. (2009), *Ecology and Popular Film*, Albany: State University of New York Press.

Stiegler, B. (2004), *De la misère symbolique 1. L´époque hyperindustrielle*, Paris: Editions Galilée.

Weber, M. (1967), *L´étique protestante et l´esprit du capitalisme*, Paris: Presses Pocket.

Notes

1   On [http://www.home-2009.com/us/index.html], acessed 20 February 2012.
2   *Earthlings* can be viewed online [http://www.earthlings.com], accessed 22 November 2011.
3   For example, milk cows are chained inside their cells throughout the whole day, unable to move, while being subjected to the effects of pesticides and antibiotics meant to increase milk production; the daily level of poultry consumption in the United States today equals the per year records of 1930, and the largest production companies slaughter more than 8.5 million birds a week; to prevent feather-pecking and cannibalism amidst the piled up specimens, which feel both frustrated and stressed, the beak of each chicken is violently burnt, and the ones of chicks are cut mechanically, at a rate of 15 birds a minute, something that often results in serious injuries for the animals.
4   Mention of Red Peter [quoting Costello], the educated primate (ape) who talks about the story of his life, his transformation from non-human animal to something approaching the human, at the Academy of Sciences, in the short story 'A Report to an Academy' (1917) by Franz Kafka. This is the tale that Costello recalls during her lecture.
5   Despite the absence of participation of national and international systems in creating regulations that would prevent and punish abuse and criminal behaviour carried out on non-human animals at the local and global levels – with notable instances in developed countries who are the major consumers – individual actions are able to manifest and reveal its impact. In the specific case of Monson, the project that began in 1999 and whose theme was the sterilization and neutering of pets, then became, over the shock of confrontation with the situations witnessed, this feature-length film.
6   The Animal Aid site reports 'As a result of our investigations, one slaughterhouse lost its supermarket contract, another has closed permanently, and legal action has either been taken or is underway against nine workers and four slaughterhouse operators. (…) And it's not just "mainstream" slaughterhouses where serious animal welfare failings were documented. In October 2009, we secretly filmed at Tom Lang Ltd, an organic Soil Association-approved abattoir that kills pigs and sheep. Conditions there were so bad that the Meat Hygiene Service immediately suspended three workers and began building a case for a prosecution. At this slaughterhouse, we filmed pigs being kicked in the face, sheep being picked up and thrown, and the heads of sheep being cut off before the statutory time had elapsed – and while they were, in all probability, still alive.' in [http://www.animalaid.org.uk/h/n/CAMPAIGNS/slaughter/ALL///], accessed January 23, 2011.
7   Quoted in Paul Cronin (ed.), *Herzog on Herzog* (London: Faber and Faber, 2002), 136. Herzog rehearses this speech almost verbatim in his autobiographical short, *Portrait Werner Herzog* (1986).

PART III

Popular Film and Ecology

China Has a Natural Environment, Too!: Consumerist and Ideological Ecoimaginaries in the Cinema of Feng Xiaogang

Corrado Neri

Introduction: China, Nature and the Second Modernization

When discussing representations of nature on the Chinese cinematic screen, scholars and the Western public tend to underline the 'rebellious' attitude of the few internationally renowned contemporary directors. Via their films, auteurs like Wang Bing, Zhao Liang or Jia Zhangke—among others—rise to the public consciousness to highlight the environmental problems that come with the dramatic changes in the Chinese economy. It is undeniably true that some Chinese film directors have indeed fostered a specific, subtle discourse on nature and ecological development, and that they use their films to elevate an ecological consciousness in the local public as well as describe the fragile beauty of Chinese landscapes by transmitting the memory of its threatened magnificence. But in the case of the majority of the contemporary market-oriented film-makers, nature assumes the allure of a product to be consumed, an avatar for nationalistic ideology, or else it assumes the place of the national identity consolidator par excellence, the 'external' enemy/the threat against which the nation-people will fight together.

In a recent collection of articles entitled *Chinese Ecocinema* (Lu and Mi 2009), most of the contributors do indeed utilize 'subversive' or challenging films to analyse the ecoattitudes of Chinese film-makers. Commentators focus their attention on films by the likes of Jia Zhangke or Lou Ye that demonstrate a critique of the government's economic strategies, especially concerning environmental politics. These authors, among others, fight with censors by providing a continuous balance of underground and commercial films, as they produce films that can be read through an ecocritical lens. Representations of nature are also employed for more vulgar (in an etymological sense: popular) agendas, which I focus on here. The scope of this chapter is not to criticize or diminish the importance of the over-mentioned 'engaged' directors, who develop compelling discourses about progress and capitalism, ecology and consumerism. Rather, I would like to focus my attention on another aspect of contemporary Chinese cinema, the commercial side and the blockbusters that try to compete with Hollywood supremacy on its own terms by creating and sustaining a star system, genre-oriented escapist movies, big budgets, special effects and the like. I focus my attention on one of the most important commercial directors of this new generation of commercial film-makers, Feng Xiaogang. I will discuss his strategy for representing nature, arguing that we can find a sensible shift from the concerns of the Fifth Generation 'return' to nature, seen as the cradle of classic aesthetics and sensibility as well as a traditional form to encode a political discourse via symbolic language; instead, Feng seems to me a perfect product of the contemporary capitalist China, transforming and using images of nature for a completely different agenda—both an industrial appropriation where nature becomes a luxury object in a material world, as well as a federating and nation-binding Other.

The Return of Nature

Already during the 1980s—the opening of China—a wave of directors and cinematographers seemed more in phase with ecofriendly aesthetics and sensibilities than with the forthcoming capitalist pulse and/or the ideology-laden Maoist past: I am referring to the groundbreaking first period of Fifth Generation film directors. Jean François Billeter (2006) points out that, from an ethnographic or religious point of view, Chinese culture has traditionally always been a centripetal one, where peripheral culture (the civilizations perceived as far from the political centre) is, and has always been, ignored. The official histories were written by state-funded Confucian scholars. As a consequence, the population does not appear in its real diversity, but only as a population subject to the Emperor—and that was enough to define them. Revolutionaries of the twentieth century transformed this administrative definition into an ethnical or racial definition. The *Han* ethnic group provided the founding of nationalism as the socialist state completed the idea of ethno-nationalist hegemony with the idea of national minorities.

Describing the return of interest in nature at the beginning of the 1980s, Billeter states:

During the 1980 a wide-spread need to return to origins and rediscover the traditions of small-town China emerged in cities. Writers and film directors expressed this in different forms. However, they were too distant from this China and ignorant about it to be able to evoke anything other than a 'dreamed China' [...]. Most often, they settled for an idealization of the image that the regime had given them of peasant immaturity, conferring it a form of absurd timeless grandeur.

(2006, my translation)

The ignorant curiosity transformed these representations into a 'dreamed tradition', where nature was described through an absurd intemporal majesty. Films like *Huang tudi/Yellow*

Earth (Chen Kaige, 1985), *Hong gaoliang/The Red Sorgum* (Zhang Yimou, 1987), or *Dao Ma zei/The Horse Thief* (Tian Zhuangzhuang, 1988) raised problems apparently forgotten in the Cultural Revolution era. They present a reflection on the mythical 'sources' of Chinese culture, history and tradition; they develop a utopian retrospective exploration of the essence and past of China—both as an imagined intellectual construction, as well as a site of wild nature indomitable by men. It is a period of redemption after the excesses of the Cultural Revolution: images of nature, deserts, rivers, colours and nuances, details and spaces (the *Vide et plein* that François Cheng set as a pillar of pictorial imagination in China, 1979) abound in these films that tend to give predominance to images instead of discourse.

This may appear anodyne to the contemporary audience but it was far from obvious during a decade that was still recuperating from the shock of Maoism. If everything is politics, if the accent is always put on the human achievements and class struggle, an image of a tree standing alone on scorched earth (*Yellow Earth*), or an enigmatic landscape barely visible in the mist (*Haizi wang/King of Children*, Chen Kaige, 1989) can suggest the presence of something invisible, utterly unsayable, open to interpretation. The return of nature, as beauty, symbolic force, chromatic palette, and the obsession of directors who have been sent to the countryside to learn from the working class, has different meanings according to different directors and their publics: a return of a religious meditation with a Taoist undertext, a search for pure, apolitical beauty, a meditation on forces stronger than humans, a non-ideological symbol. Or else, at least, something—an image, a suggestion, a hint, a symbol—capable of escaping the control of the party. But since all images and/or stories can be ideologically driven or become a vehicle for embedded ideology, the shifts in the cultural paradigm let the images of nature be progressively appropriated both by critics of the regime as well as its spokespersons—and this came as an important change in China, where ideology used to be (and still is, of course, in some of the most explicit propaganda-driven films) clear-cut and clearly identified.

The search for roots of Chinese intellectuals in the beginning of the 1980s, when passing through natural imaginary, has had the important function of both giving pre-eminence to art as a free and plurisignificant reality, as well as to inscribe human beings in a philosophical, metaphysical or emotive context much bigger than their political or ideological struggles. The representational tradition of landscape literati brush painting is an evident anchor to link nature, tradition, and the possibility of a symbolic art that develops an implicit discourse comprehensible only to the educated few, delivering at the same time an aesthetic pleasure of a more trans-class nature. The literary trend of the search for roots was accompanied by the 'wound' literature, an (auto)biographical trend that traced the suffering of people during the Cultural Revolution. This trend of engagement spurred later directors like Jia Zhangke or Lou Ye to, more or less undercover and more or less independently, produce images that can be interpreted as harsh criticism of the regime, or at the very least a cry for regaining the ecological consciousness that has been sacrificed at the altar of economic development.

The most representative examples are *Sanxia haoren/Still Life* (Jia Zhangke, 2006) and *Suzhou he/Suzhou River* (Lou Ye, 2000). The first one is an elegy of the Three Gorges landscape, destined to be submerged after the construction of the titanic dam—one sequence tellingly contrasts the landscape as it appears printed on the *renminbi* banknotes against the background of the actual contemporary landscape disfigured by works and building sites. *Suzhou River* is a modern noir set against the background of Shanghai, described as a polluted grey town, its river dirty and smelly. If these films are particularly representative, they are by no means unique: other directors, like Wang Bing, Ning Hao, Wang Xiaoshuai, Zhang Yang, Huo Jianqi and others have shown an environmental conscience and an acute attention to ecological problems in their work, denouncing human greed that destroys natural resources, thought shocking images of pollution, waste of natural resources, or human sickness due to exposure to industrial venom.

Real Nature and the Politics of Globalization

In China too, we can, surely, see a contrast between an ecoconscience—that can acquire the dangerous nuance of a critique of the regime's politics—and the drives of rampant capitalism. An ecoawareness is *de rigueur* for contemporary artists: omnipresent in cultural products, advertising and television, the aesthetics of a return to nature stems from a double source. *In primis*, the alleged tradition of Chinese art, arguably in a holistic relationship with the forces of nature, is remixed for the new millennium in a sexy and vaguely New Age frame (yin yang, traditional medicine/homeopathic treatments, five elements and reincarnation, colour symbols and pleasure of gardens). The other source is the globalized frenzy for bio-products, pollution awareness, anxiety concerning the threat of global warming, the globalized fad of agritourism, and generally speaking, the package of nature–sexy integralism wrapped up in a mixture of bio-consumption, all keywords imposed by spin doctors of various multinationals.

While China is one of the most polluted places on earth, especially in the cities, and various scandals related to contaminated 'natural' products periodically cover the front pages of newspapers, popular cinema presents nature as an image to be appreciated or as a symbol of national grandeur rather than a treasure that has to be protected. Nature is a selling tool for all kinds of products, designed ultimately for upper class leisure. If periodically the state censors or tries to reduce the influence of the 'Western' pernicious mentality, for example with campaigns against luxury goods and their magnification, it is evident that status and privilege are the drive *par excellence* of the recent and virulent capitalist frenzy as wealthy people drive silent sport cars through valleys and deserts, play golf in uncontaminated natural landscapes, eat and put on their skin the most refined natural fruits—and their by-products. The danger inside this specific marketing strategy is, like in the West, the phenomenon of 'ecowashing': behind the facade of selling ecofriendly products and donating a percentage of the benefits to ecological projects, lurks a marketing strategy that is, moreover, more expensive (and ultimately damaging) than the cost of ecofriendly action.

Feng Xiaogang's films perfectly describe the shift of cultural paradigms in contemporary China, and his representation of nature enables us to analyse the dominant discourse on nature sold in media-China. Feng is by no means the only market-oriented Chinese director, but he is probably the most representative of the shift in the politics of production, distribution and consumption of cinema in contemporary China. It is from this that the definition, sketchy and imprecise, of the 'Chinese Spielberg' emerges, a notion connected also to a recurring desire to paragon and equate Chinese directors/actors (and cities—Suzhou/ Venice, Shanghai/Paris) to a Western equivalent. Feng is indeed both a commercial success and a daring explorer of different genres, a star-system-based producer and a middle-class poet, at ease both with huge production as well as with middle-brow urban films.

Feng started his career in the mid-1990s with the production of urban comedies aimed at local audiences. Surfing on the wave of fast commercialization of the Chinese market, and benefitting from the help of the government to increase the local affluence of film theatres, Feng proved his clever businessman skills, always keeping both an eye on the public and on new techniques of globalized—some might say, 'westernized'—cinematic production. Feng's films show how deep the influence of Hollywood, perceived both as a rival and a model, runs in China. If the Fifth Generation films could claim a return or rediscovery of ancient traditional culture stemming from historical visual representation, from painting to New Year's woodprints, contemporary Chinese blockbusters are formally indebted to Hollywood standards: invisible editing, Wagnerian score, construction of plots according to classic Hollywood narrative structures, star system überpower and the like. In addition, Feng's films are also representative of a new vision of popular entertainment: aggressive marketing displaying sexy protagonists as idealized, purified beauties, product placement, strategic release dates, and the conventions of 'high concept' product conveyed in a few lines. While his films have never really sold well in foreign markets, Chinese audiences are enthusiastic supporters of his products.

Feng shifts from comedies and satire (*Bujian busan/Be There or Be Square* 1994, *Shouji/Cell Phone* 2003) to action (*Tianxia wuzei/A World without Thieves*, 2004), from war (*Jijie hao/The Assembly* 2007) to costume drama (*Yeyan/The Banquet* 2006), from historical films (*Tangshan da dizhen/Aftershock* 2010) to sentimental comedies again (*Feichang wurao/If You Are the One 1&2* 2008 and 2010). He is a highly capable actors' director, having worked with stars like Zhang Ziyi, Andy Lau, Ge You, Shu Qi, as well as a daring risk-taker—often switching genres and styles. In this sense, he embodies a 'new', globalized but nationalistic China: this is a self-confident industry, not in need of a Western market for its cultural products (but aiming at pan-Asian visibility). Entertainment-oriented mass production combines here with embedded ideology far from explicit propaganda, as chauvinistic messages work alongside product placement in a vertiginous mingling of the market economy and state controlled art.

Feng's work does not strictly deal with environmental issues, nor does it focus explicitly on transnational connectivity, with few exceptions like *Da wan/Big Shot's Funeral* (2001), starring Donald Sutherland. However, this is precisely the reason they provide a privileged perspective through which to analyse the discourses and the politics of representation of nature. His environmental politics tells us of the ways the ruling class in China disseminates a message of blind consumerism and wider globalization—intended here as an assimilation and acceptance of the transnational capitalist and consumerist system, even if it is 'protected' by a discourse that stresses a nationalistic chauvinism.

Paradoxically, the globalization of China seems in fact, to me, always deeply nationalistic and, precisely because it is nationalistic, widely globalized. I use the term 'paradoxically' because, while capitals, modes of production and consumption, lifestyles, architectures, clothes, etc. are becoming more and more similar worldwide, politics emphasize the nationalistic framework to protect their unity and economics. This 'empty' (adaptable) nationalism seems the only ideology left, to the point that it could be described as post-ideological. This empty ideology looks more and more elusive and neutral: the clear cut, socialist emphasis on the international agenda of proletarian forces fades out in favour of a transnational and adaptable quest for happiness and/or harmony—strangely resonating both with state dirigisme and the American constitution. No longer concerned with the communist agenda against the evil forces of capitalism, the ethnic-nationalistic entity of the State adopts the economical and—arguably—ethical rules of the ancient 'imperialist power', while it stresses (and invents) its specificities. In this post-ideological environment, in a context where religion is an everyday practice of apotropaic 'superstition' (as the red guards would have put it) but not an ideology, the most successful Chinese films at the local box office construct a rhetoric of return to nature that does not hide very deeply the superficiality of interest in environmental issues.

In my view, Feng's films are the perfect incarnation of how the image of uncontaminated nature is used in a capitalist and consumerist world to sell products, or to be the correlative objective of human feelings, and finally, when nature is described in its most wild and dangerous forms, as a catastrophic force requiring a *political* reaction—a nation-builder crisis. We are no more in a Fifth Generation context where the plurisemic allegories described by images of nature helped a (re)newed consciousness of the incommensurable depth of art, reflecting on cycles of life and mysticism, death and beauty. Even if many images of nature are presented in Feng's film, we are not in a Jia Zhangke or Wang Bing film, where viewers are requested to position themselves in a critical stance. We enter into an ultraliberal post-ideological entertainment society that communicates through entertainment its political (nationalist) directives. As the guidelines of the Party evoke the 'harmony' of the society and, recently (2011), the legitimation of the search for 'happiness' that Feng's films compare with the Western postmodern legitimation of consumption—a strategy that aims at the creation of a more equilibrated society where more people would be able to consume and buy and reach the 'middle class' status.

To discuss these wider transformations, I focus on images of nature from Feng's films that in my view hide a political agenda, that are embedded with state's strategies, and that are used in a specific poetical/artistic way. The order is chronological—even if I do not think there is an 'evolution' in Feng's discourse around nature. Since these films are all successful and trendsetting, it seems important to read the implicit agenda and ideological representation of nature that they subsume as relevant trends of entertainment cinema in contemporary China.

Nature as Correlative Objective

A World Without Thieves employs the natural image of Tibetan mountains as a symbol of searching for oneself and introspection. The story follows protagonists Andy Lau and Rene Liu on a long train journey, where they meet an innocent boy who is threatened by a gang led by Ge You. She is, as expected, pregnant and wants to stop leading the life of a thief to grant the yet unborn child a respectable, honest life. This determination to purify her life, helping both the unknown boy and her man, finds an immediate echo in the images of Tibet. These natural images—mountain peaks, floating prayer flags against the breathtaking scenery of a greenish valley surrounded by rocks—are immediately inscribed in the human context: praying people dressed in the traditional Tibetan garments contrast with shots of imposing ancient temples that seem almost part of the natural landscape and earth—in terms of colour and mass, the temples are inscribed in nature as if they were a component of it.

The ideological discourse could not be more obvious: Tibet is represented as a pure land of Buddhism, an uncontaminated place where humans are in harmony with nature, where the craziness of global, modern, urban life has not yet arrived—it is indeed a cliché, but omnipresent in the Chinese cultural sphere, which is even more striking if compared with the silence around political/social Tibetan issues. The roaring car driven by the two protagonists makes a stark contrast with the peace surrounding them, emphasizing the starting point of the story: the two thieves are no longer in harmony with one another and with life. They need to accomplish a long journey, both physical and spiritual, to attain illumination and renounce stealing—this redemptive ending resonates with Feng's ambiguous poetics always in balance between the exaltation and criticism of the new upper class: the films use satire to depict the nouveaux riche while their elaborate staging of the life of this new upper class contributes to the creation of a desired model of a-ideological wealth. Furthermore the long train journey, which is the main section of the film, symbolically links Tibet to China, abiding by the official dictation and setting aside the political troubles that regularly hit the region.

Tibet is an already packed symbol of redemption, of return to sources of spirituality, and purification from earthly desires. As previously discussed, many Fifth Generation directors have already explored the 'savage' nature of the Chinese landscape, including deserts and high mountains, forests and barren landscapes. These images remind us of the reeducation to be carried out far from the cities, in the countryside, and of the discovery of a primitive culture (in the sense described by Rey Chow 1995) that is part of the canon of representations of Chinese civilization. For the urban city dwellers, these images of Tibet (seen so often in television news and advertisements) represent a defamiliarization, a way to escape from daily life and eroticize/sensualize their 'own' national culture. There is no questioning Tibet's belonging to China, no mention of political tensions in the region. No politics, just nature—as if it were possible!

Images of nature inserted in a film are a way to contextualize the plot, but also to call out for the sensibility of the spectator to the representational *doxa* of tradition, especially in the form of landscape painting that was one of the most practiced, rated and symbolically charged art forms in China. Many directors work on the (vague) symbolism of colours and their interrelations with the theory of the five elements and the holistic conception of the universe—see, for example, the discussion on the semantics of colours in Zhang Yimou's *Yingxiong/Hero* (2002) and his technique of filming as creating *yijing*, a nebulous but evocative concept taken from classic painting. *Yijing* means 'idea-image':

> The power of the concept lies in the acknowledgement that neither the cinematic nor painterly image is real. It is an idea-image that is not only rendered visible on-screen and improved by digital technologies: the image also carries an introspective aura, an emotional register, and a cultural heritage. It reveals a landscape of the heart and mind.
>
> (Farquhar 2009: 100)

The concept of 'idea-image' can be used for the images of nature that pop up in chromatic magnificence in Feng's *A World Without Thieves*, helping to tell the stories behind the main plot, reactivating in the viewer suggestions and correspondences: the mountain and valleys that the train crosses immediately situate the journey in what is perceived as a traditional landscape, that brings the viewer to a context of transcendence, of hermitage, of spiritual search and inner journey. In Feng's film, as in many other contemporary Chinese mainstream films, nature representation is an aesthetic value and a form of control—intimate and nationalistic control of wilderness that creates an idea of national identity as constituted by a vast, omnicomprehensive variety: outside the city, there is a mystical force that is part of *our* nation. Discussing *Shijie/The World* (Jia Zhangke, 2004), Silbergeld argues:

> As in conjuring, voodoo, and Chinese alchemy, the role of replication, miniaturization, and substitution, are tied to the achievement of mastery, the accumulation of power, and for the individual, clan or dynasty especially, the attainment of longevity. The Qin emperor Shihuangdi, the Han emperor Wudi, the Qing emperor Qianlong all had the same thing in mind when constructing their parks and that thing is, in a word, control.
>
> (2009: 126)

This idea of control is reminiscent of Jean-Luc Godard's monumental *Histoire(s) du cinéma*. Its fourth chapter is called 'the control of universe', and it states that the great directors achieved what Napoleon and Hitler never could, the control of (their) universe. Stretching to Feng Xiaogang, and combining the idea of control to the concept of idea-image, we can see how his representation of nature is far from any environmental politics, while approaching at the same time the control of his audience—manipulation of emotion and transmission of hidden ideology—by the figurative control of nature through the illumination achieved by his characters.

This does not appear to me as unique to Feng, but part of a vast international New-Age wave. The oeuvre of Terrence Malick heavily utilizes grandiose images of (National

Geographic-like) nature to contextualize human suffering in a wider metaphysical scheme, especially in *The Tree of Life* (2011), but also in *The Thin Red Line* (1998 and *The New World* (2005). Lars Von Trier (especially in *Antichrist*, 2009 and *Melancholia*, 2011) constructs lyrical plastic symphonies where nature is represented as a sublime, mysterious and terrible correlative objective of humanity's (specifically: feminine) deeply rooted anguishes and Secrets. Likewise, we could think of other films like *127 Hours* (Danny Boyle, 2010) or *Into the Wild* (Sean Penn, 2007) where nature manifests a hard facade, but it also helps the solitary human being to escape civilization and seek for hidden truths and forgotten wisdom. 'Primitive' wisdom is to be found in Pandora, where natives live in a USB-like plug-in with Mother Nature (*Avatar*, James Cameron, 2009); or 'ancient', pristine nature is both a new beginning (completely uncontaminated and unspoiled) and awe-inspiring (dinosaurs and fierce beast running wild) in the TV show *Terra Nova* (Silverstein and Marcel, Fox, 2011).

Such depictions prevail in the Chinese context as well, where films like *Guanyin shan/ Buddha Mountain* (Li Yu, 2010) connects personal suffering (the mourning of a son/ daughter) to the national tragedy of the great 2008 Sichuan earthquake. In *Buddha Mountain*, the protagonists—a bunch of stylish dishevelled and smoking Calvin Klein-like young idols and the star Sylvia Chang—help rebuild a temple in a green forest on a mountain as they manage to come to terms with personal grief and questions about life. Again, the film draws a path of redemption. Nature is a vehicle for enlightenment and wisdom: the last images show Sylvia Chang on a cliff, and then she disappears. We will never know if she committed suicide or not (she tried already, failing), but the message is that she merged with the cycle of nature, which can be pitiless, but can also elevate the human soul to a higher level of consciousness.

Nature as Status Symbol

Nature is depicted as a commodity in the two instalments of Feng's megahit *If You Are the One*, a romantic comedy where the stars Shu Qi and Ge You indefinitely postponing their marriage ceremony. The stunning landscapes in *If You Are the One* remind us of the importance of nature, bucolic surroundings and the absence of human intervention within the changing of seasons, hinting at the same time at the dangers of losing such natural richness. Simultaneously these 'built-up' symbolic images refer to dreams of capitalist possession where nature turns into nothing but a marketing tool, a status symbol, a showcase of wealth and vanity parade.

Ge You plays a hyper-rich contemporary businessman who embodies a critique of the capitalist society, its tics and neuroses, while he indulges in every possible pleasure and luxury, lives in villas mounted as gems in a majestic forest, travels in private jets and attends the most exclusive social events. The representation of nature in these films is at the same time omnipresent and strangely elusive. During their long honeymoon, the protagonists travel to astonishing regions (he is incredibly rich, she is a flight attendant), and are supposed to be in contact with nature. However, they are never alone in nature: visible screens, a window, a television, or architecture constantly protects them from nature (and possibly, protecting nature from them). For example, as the couple sip cocktails inside a hyper-stylish bungalow, a window opens to a luxuriant forest; meanwhile, natural surroundings are seen on the screen of a television (which transmits the gala Shu Qi is attending), or through the window of a luxury car or first class on a plane.

In *If You Are the One 2*, the absolute unconcern for environmental politics is even more striking: at one point, Shu Qi says that she prefers a shower because she has an environmental conscience and wants to 'protect the environment', only to be seen, five minutes later, enjoying a bubbly bath inside the resort. Still in the same resort, Ge You is shown diving into a swimming pool: the pool is filmed from a high angle as we see the clean water of the artificial swimming pool enclosed by its borders, and in the background, the immensity of the sea. But the high class is, of course, bathing in an artificial pool, comfortably framed and separated from 'real' nature. The only moment when Shu Qi goes into the ocean, she is drunk and in despair.

To me, this form of representation parallels environment to a commodity: nature is a luxury item, to be enjoyed from a secure, privileged point of view, even better when idly sipping a colourful cocktail. Nature is grafted into simulacra, where its 'realness' has to be constantly denied so as not to disrupt this illusionism. This sort of nature-becoming-merchandise enables the possibility to travel to exotic destinations, the opportunity to enjoy grandiose scenarios, and possibly let friends know that we are enjoying it. Feng is not alone in presenting nature in this way as other ironic examples can be found in the blistering *Juezhan Shama zhen/Welcome to Shama Town* (Li Weiran, 2010), where a desert region becomes, with a Zhang Yimou-like marketing operation (Zhang is openly addressed in the film), a tourist attraction.

Feng's story of the *nouveaux* riche, of social climbers, does not avoid talking about their weaknesses and vulgarity, but at the same time, such a representation, like the Bollywood standards of lush scenography and European Alps ballets, shapes the object of consumerist desire, and creates a chimera that represents the standards for which common people should crave. Feng's focus on displaying wealth is not simply a Chinese phenomenon. The showing-off of wealth is often presented as a critique in many other cultural contexts: see the Italian *cinepanettone* where every year for Christmas, a bunch of stereotypical Italians leave the reassuring grounds of the nation for exotic destinations, intended to form a postcard background for their never-ending love affairs; see the 'girls' of *Sex and the City* (Darren Star, HBO, 1998–2004) that leave New York to visit, on high heels, an exotic Arabic and Orientalized country; see the dream within the dream in (yes, another) dream in *Inception* (Christopher Nolan, 2010), where rich and beautiful protagonists virtually travel in luxurious Jamesbondian exotic and wild landscapes, from *inside* an ecounfriendly plane.

If you Are the One displays such a spectacle of nature, reflecting the show-off vulgarity of Beijing's new wealthy upper class and the need to frame and control nature. If we endorse the hypothesis of Feng as a conscious challenger of social norms, as Zhang Rui tends to do (2008), we could find here a strategy to condemn the empty, greedy and manipulative globalized television culture of appearance. Still, even if we can appreciate the irony displayed by Feng, the very representation of luxury and wealth creates its own target: the spectacle of richness attracts the mass public that 'needs' to dream of an unattainable life. This sort of *mise en spectacle* helps create a greedy, capitalist-oriented, ideological and remissive society that feigns to despise the high class, but ultimately models itself on the high-class illusion, its way of dressing, of loving, of talking, of travelling, of participating in politics and social life. These films are far from being neutral: they indirectly manufacture a social ideal and an aesthetic value—not by chance, modelled on an anonymous, transcultural, globalized oligarchy. Again, *If You Are the One 1&2* are 'just' comedies—there is even a sequence where Feng mocks the Chinese underdeveloped environmental awareness (he stages a ridiculous TV-gala event to raise funds to protect penguins): yet, these films clearly embody the national pride and the political direction advocated by the party.

To take this argument further, Feng's films are also known for the extent of their product placement, a connection that creates a close link between the consumerist politics of a privileged and 'natural' tourism and the liberal circles of the Chinese society. *If You Are the One* in particular had already recuperated its initial budget before even being screened, thanks to the commercial strategy of placing indirect advertisement in key sequences. Nature itself becomes an actor that advertises all sorts of products, as it becomes an object of consumption, to be preserved, yes, but even more to be possessed, framed, displaced— symbolically, ideologically, materially. Emmanuel Paris puts forward an interesting reading of the contemporary creation of the myth of a remote, savage, spectacular nature by the creative industries:

Turning nature into spectacle is [...] the means for the creative class to be recognized as guarantor of the correct social order, and arbitrator of individual and collective life in the form it should exist in. [...]. Around the world, virgin territories are also redesigned following such good practices; 'ethical tourism' is a burden for the lives of the populations of the Amazons or the Nepal mountain regions. The tourist appetite for these populations and territories is so strong that the creative industries somehow perpetuate the rumor of a persisting wild state, and of naturally spectacular nature. The *mise en spectacle* of nature by the creative class makes in other words invisible those cultural practices carried out by human societies living in natural spaces – including the most traditional and ancient ones.

(2010: 26–27, my translation)

What Paris points out is the importance of the creative class, of the spin doctors, of the image of the new, globalized world in which China in many aspects resembles the Hollywood image of the West. Worldwide, upper class is never in touch with nature, and never really concerned by environmental politics—which would imply a political/social engagement;

but, via their luxury travels, galas, golf playing and organic restaurants and the like, they are always artificially immersed in the construction of 'wild' uncontaminated nature.

As a counter example, we could point at the pictorial work of Yang Yongliang: the painter draws huge landscapes that, from a distance, resemble the traditional, monochrome ink brush literati compositions. But if observed closely, it is possible to see that the texture of the mountain, rivers and forests is no longer organic or biological, but has been substituted by a web of machines, excavators, cables and cranes. Here the illusion of uncontaminated nature is mirrored in deformed glass and reveals the ecological catastrophe of the contemporary world. In Feng's films we can see a nature that seems to be there to please the exotic desires of the new Chinese upper class; we cannot see the ruins or the ecological problems, but only a magnificent representation where the pride of Chinese economics can be staged. On the contrary, and similarly with the works of Jia Zhangke or Wang Bing, in the pictorial work of Yang Yongliang we can perceive, *behind* a *trompe l'oeil* 'traditional' landscape, the disasters and the destructions caused by rapid industrial growth, too often disrespectful of ecosustainability— and there is no 'reading between the lines' here, but an explicit, foregrounded stance.

Nature as Enemy

Finally, we turn to *Aftershock,* where blatant nationalism and entertainment mingle together more explicitly. Initially, the film was supposed to be dedicated to the memory of the victims of the devastating earthquake that in 1976 destroyed the city of Tangshan, the same year Mao Zedong passed away. The mourning for the victims of Tangshan was forgotten in favour of bereavement of the entire nation for their leader. Superstitiously, it was also possible to see the awakening of Mother Earth as a sign that the heavenly mandate was over for the Communist dynasty—and of course, rumours of that sort were put into silence. During the shooting of the disaster film, another catastrophe occurred, the huge earthquake in Sichuan in 2008. Feng feared that the project would come to an end, but instead he managed to link the two stories and create one of the most successful Chinese films ever.

A family is destroyed by the 1976 earthquake; the father dies, and the mother has to choose which one of her children to save, the girl or the boy. She chooses the latter and spends all her life in remorse. The boy grows up and, even if he lost an arm during the earthquake, he embodies the young China able to make money and ensure a pathway towards modernization, capitalism and wealth. The girl, unsurprisingly, was not dead, but lived with the memory of her mother choosing her brother over her. She marries a rich Canadian and emigrates to the West; but when she sees on the television the images of the terrible 2008 earthquake, she comes back to China, finds her brother, and reconciles with her mother in a deluge of tears.

As Brice Pedroletti (2010) puts it, catastrophe films are at their most successful when displaying the whole country ravaged by natural disasters. *Aftershock* touches an open nerve, an extremely sensible point in the Chinese consciousness. As we can infer, all 'natural' disasters can be more or less devastating according to how humanity manages their coexistence with nature—indiscriminate constructions, nuclear plants built on tectonic plates and the like do not create natural disasters, but increase their potential dramatic impact on humanity and nature. This film does not ever take into consideration the idea of criticizing or questioning the sanity of the constructions (notably, of the schools that went down as if they were made of clay) nor the possibility of a sustainable urban development. It does, in contrast, create and enhance a strong feeling of national identity, of national pride and collective mourning. The Tangshan earthquake, so near the death of the Great Helmsman, was silenced by the government. Now, times have changed and public mourning is a part of national identity construction. The authorities handled the catastrophe very well: they were very fast to run to the disaster zones, the prime minister was filmed cask on his head bringing comfort to the victims, television and institutions organized countless fund-raising events and ceremonies to help commemorate the victims. In another context, we could think of the political documentary *Draquila* (2010), by the Italian film-maker Sabina Guzzanti, which tells about how the L'Aquila earthquake was a god-send chance for the Prime Minister Berlusconi. Suffocating under the pressure of several scandals, Berlusconi could use the catastrophe to mould him as a saviour and shift attention away from the scandals. Indeed, 'the exercise of mourning has become an ardent obligation in the case of catastrophe in China' (Pedroletti 2010; my translation), where all media are summoned to participate in the nationwide catharsis.

In *Aftershock*, human responsibility (no security norms, no urban plans, fast-paced progress that tears apart natural rhythms, etc.) is never confronted. Humans are all good-hearted people, ready to relieve victims and help the innocent. Tragedies, it is well known, unite nations and help construct a sense of imagined community, especially in the present-day mediatic world where communication and representation are the vehicles of ideology. National sentiment is inflamed, piety unites and gathers the citizens, all critiques are easily silenced by the necessity to help and reconstruct.

Discussing the Hollywood catastrophe film *2012* (Roland Emmerich, 2009), Charles-Antoine Courcoux points out how:

Cataclysm is a sharing force through which the narrative distinguishes, within the framework of beneficial rebalancing, between what must disappear and what can be saved. [...] On the level of locations and objects, the film contrasts the destruction of the sites that, in the collective imaginary of the United States, are considered the quintessence of greed and materialism (Los Angeles, Las Vegas, a supermarket, sports cars, a luxury cruise), with the destruction of heavenly nature (Yellowstone and Hawaii) which carry basic values of this culture. This way the narrative pushes us to understand that consumerism, covetousness or technological dependence are never controllable nor limited practices, but instead a wide-spread threat that eludes its instigators, a disposition that affects even the foundations of a strong, prosperous and virtuous society.

(2010: 162–63, my translation)

However, in *Aftershock* the two earthquakes do not represent a critique of over-capitalism nor of destruction of nature's rhythm. There are no more enemies to fight, class struggle has been achieved, the United States and the West represent economic rivals at the most—except when periodically a US president accepts a visit by the Dalai Lama, which unfailingly provokes harsh reaction from the Chinese government. The construction of the enemy is one of the very bases of nationalism and national identity: here nature plays very well the role of a fierce, unpredictable enemy. Human society is unable to fight with nature, but can evoke and stimulate the human values that help it to get stronger and overcome difficulties— and indeed we are shown that is exactly what is happening in China. It is about catastrophe politics, and an efficient use of human needs to unite and help each other. This political vision of *Aftershock* does not want to be unrespectful of the real sufferings of the victims and their families. It is important to stress that, in a very direct and unsubtle way, the film embodies a nationalism that touched the heart of the public, and contributed to guide the population according to the keyword of the Party: 'harmony'. We can see how Feng's film manages to wrap up important ideologies, conveyed through tears and conventions of high-concept blockbusters.

Courcoux underlines another characteristic of the Hollywood disaster film that also remains very different from its Chinese epigone. Through a close analysis of *2012*, Courcoux shows how the movie gives shape to a mythology of individual reconstruction, and in particular, a reconstruction of masculinity:

> If the natural catastrophes put on screen by *2012* draw on an environmental construct, it is that of a form of ecomasculinity, of a nature that, in times of technological development and social change, is led to play an elective and essentialist role to remind us of the innate superiority of the white middle-class man, and simultaneously underscore the need of making him into the figurehead within a new transnational order.
>
> (2010: 168, my translation)

This 'hegemonic masculinity' is not present in the Chinese context, or at least not in these contemporary blockbusters. Unsurprisingly, the hegemony that can save the world and society in a holistic embrace is family. The tragedy occurs, in the 1976 earthquake, because the mother has to choose; and any wealth and social status that her son could give her are not enough, not until she finally has the chance to ask forgiveness from her daughter and recompose the mutilated family. From a Confucian perspective, this family-centred ideology is both personal and political: the greater family is the Nation, the equilibrium of the nation stems from harmony in the family, the 'rectification of the names' inside the familial context (the father must be the father, brother must play the role of brother and so on) will eventually lead to the harmony of the virtuous society. While family is, of course, a pivotal theme in Hollywood and western cinema, what is striking about contemporary Chinese discourse is that the Confucian style family/ideology is perceived and described as *specific* to China—it is the base and source of national unity, strength, stability and, possibly, superiority. Adaptation of the traditional Chinese family to suffering is a token of political stability, social harmony, and civic strength even when facing catastrophe—which cynically can and is instrumentally used not to reflect upon environmental issues, but to reignite the fable of a strong China, of the unity of the people, and the mythical progress of the Chinese civilization.

As forms of symbolic power imposed on the collective imagination, and symptoms of the ever-growing need to be aware of the fragility of nature and the dangers of uncontrolled development, these films, if read 'against the grain', bring into play several contradictory discourses that cannot be ignored in the future development of China. Like every great producer of popular cinema loved by their audience, Feng Xiaogang produces narratives governed by the populist and consumerist *doxa*, from which potentially subversive—or, to say the least, reformist—apprehensions emerge. After the tears and the catharsis, a viewer may want to ask further questions, which are left unanswered by the film. These include: why such a long silence after the 1976 earthquake, and why such a public display of grief in 2008? What happened to the school that fell down killing hundreds while government buildings are still standing? Is there something that government and politicians could do, if not to prevent a natural catastrophe, at least to minimize its disasters? Representations with an environmental conscience that can be in tune with a severe authority controlling media discourse will have to find a subtle, and presumably necessary, balance between growth and sustainability.

Consuming Nature and Representing Ecology

Feng Xiaogang's cinema is not intentionally ecologist. It does not provide a denunciation of human exploitation over nature, nor accuse the rapid economic growth as a source of global pollution and warming. Feng's works are not explicitly environmental friendly, nor do they advocate a return to nature or a deeper respect for its rhythms. They do, however, implicitly address the contradictions and different anxieties linked to China's economic and environmental development. In the background, these narratives give away several discourses and policies more or less connected to the theme of nature and its difficult relationship with a society that is undergoing unprecedented economic development.

We could attempt here a summary classification—simplified, indeed, but one that tends to demonstrate the ultimately global nature of Feng's representation of nature—indicating that the discourse of Chinese box-office hits has often been moulded by Hollywood standards going global, with the recent Chinese blockbusters both resisting and assimilating them. First of all, nature is represented as a mystical/mythical other. Nature is what globalized, urbanized, transcultural population lacks: a distant image of spirituality, of roots, or enlightenment (*A World without Thieves*). These features are often attributed to a critique of capitalism, because the capitalist mode of production is regularly stigmatized as unrespectful to nature, since it privileges conspicuous consumption regardless of the devastating effects that industrialization can bring to the ecosystem. Feng is both a critic and a spokesperson of the new upper class. In his films, we find both a critique and praise of the rich people's way of life and consumption—an attitude intriguingly contradictory but also commonly globalized.

Second, nature can be perceived as a form of luxury, as we see in *If You Are the One*. The proletarian class (a nasty incorrect word, for sure, especially in one-child policy China, but still appropriate in my view) lives in a modernized, industrial megalopolis, while the upper class can afford to inhabit ecofriendly cottages in the hills, eat organic food, and to visit remote uncontaminated settings. By using these vague categories, I intend to point out that we can witness a growing disparity between the poor and the rich, between the *mingong/* irregular workers coming from countryside to build up the megalopolis' skyscrapers and the ruling class, the happy few who can afford to live a luxurious and healthy life. This is by no means specific to China as it is also happening in other Asian countries as well as in the West, but public awareness about scandals such as tainted milk and unequal access to high grade food supplies is increasing in both the official channels and the Internet-based popular discourse. While Feng's films offer a plain critique of vulgar consumption and obvious social disparities, simultaneously, the alluring representations of the nouveau riche way of life creates a desire, a model, a normative implicit encouragement to get rich, and to get rich as fast as possible.

The landscapes in Feng's films are taken over as consumer goods by a new ruling class, the seasonal cycle is modified to convey a formulaic love language, and beauty is worked upon to counter Hollywood's control over the collective imagination—until *Aftershock*, where two earthquakes take place in order to depict the image of the ultimate enemy around which to gather socially and spiritually. Here, we see the third category emerging where nature is the ultimate enemy: apolitical, federating and consensual. National identity needs an enemy and nature can incarnate a fierce power that (apparently) escapes ideology and politics.

These categories can and will overlap; they indeed represent both a specificity of a new vision of Chinese popular culture on nature and environment as well as show a national cultural system intimately interrelated with the globalized flow of production. China is integrated in the global flux of capital, and it will become more and more significant in global economic and ecological strategies. Some films explicitly denounce the unscrupulous tycoons ready to destroy natural resources and beauty only to make more money, unconcerned by ecosustainability. Feng's film tells us on an immediate level that China is becoming modern and globalized—enjoying the capitalist way of life, being more consumerist than the consumerist nations, adapting its own cultural industry to resemble the Western dominant entertainment (music, editing, genres, etc.) and to allure the outside world, Asia *in primis*.

But at the same time, with a critical lens, I argue that it is possible to see behind the spectacular and aesthetic frame and read different patterns, independently from the agenda of the director and producers: nature is an enemy—but why? As I argued before, the changes in the post-Cold War world had taken away from China, and from other countries as well, an immediate, clear-cut, stereotyped enemy. Mother Nature can be represented as the uncanny force that threatens the world as we know it, against which the community of men can and has to unite. Besides that, nature can and often is perceived as a mythical other, a force that has been forgotten by modern human communities and that must be recovered in order to survive first, but also to retrieve a mythological, ancient wisdom related to a 'primitive', original holistic union between human beings and the other inhabitants of the planet. The disaster film genre is relatively new to China—and in my view it is coming to the foreground now precisely because nature can be perceived as a post-ideological enemy—which has not been the case since the end of the 1970s. Feng's films do not nor are they intended to answer these transnational questions. But they can help to pose them—even if this is done in an indirect way.

References

Ban, W. (2009), 'Of humans and nature in documentary: the logic of capital in *West of the Tracks* and *Blind Shaft*', in S. Lu and J. Mi (eds), *Chinese Ecocinema: In the Age of Environmental Challenge*, Hong Kong: HK University Press, pp. 157–69.

Billeter, J. F. (2006), *Chine trois fois muette* (China: Three Times Speechless), Paris: Allia.

Cheng, François (1979), *Vide et plein: le langage pictural chinois* (Blank and Full: the Pictorial Language of China), Paris: Seuil.

Chow, R. (1995), *Primitive Passions: Visuality, Sexuality, Ethnography, and Contemporary Chinese Cinema*, New York: University of Columbia Press.

Courcoux, C. (2010), '*2012* ou la politique de l'écomasculinisme' (*2012* or the Politics of Ecomasculinity), *Poli*, 3, pp. 151–68.

Farquhar, M. (2009), 'The idea-image: conceptualizing landscape in recent martial arts movies', in S. Lu and J. Mi (eds), *Chinese Ecocinema: In the Age of Environmental Challenge*, Hong Kong: HK University Press, pp. 95–112.

S. Lu and J. Mi (eds), *Chinese Ecocinema: In the Age of Environmental Challenge*, Hong Kong: HK University Press

Paris, E. (2010), 'Spectacle de la nature et classe créative' (The Show of Nature and the Creative Class), *Poli*, 3, pp. 17–27.

Pedroletti, B. (2010), 'Le film qui fait pleurer les Chinois' (The Movie that Makes the Chinese Cry), *Le Monde*, August 19, 2010, pp. 77–78.

Silbergeld, J. (2009), 'Facades: the new Beijing and the unsettled ecology of Jia Zhangke's *The World*', in S. Lu and J. Mi (eds), *Chinese Ecocinema: In the Age of Environmental Challenge*, Hong Kong: HK University Press, pp. 113–27.

Zhang, R. (2008), *The Cinema of Feng Xiaogang: Commercialization and Censorship in Chinese Cinema after 1989*, Hong Kong: Hong Kong University Press.

And the Oscar Goes to ... Ecoheroines, Ecoheros and the Development of Ecothemes from *The China Syndrome* (1979) to *GasLand* (2010)

Tommy Gustafsson

W hen the Three Mile Island accident happened near Harrisburg, PA, in March 1979, it became the start of the Anti-Nuclear movement—and thereby the political Green movement—that stalled building and planning of new nuclear power plants in the Western hemisphere for decades to come (IAEA 2011: 3–6). Just over a week before the accident the drama thriller *The China Syndrome* (James Bridges, 1979) had premiered to first-rate reviews as a 'terrific thriller', even though some saw it as a biased political film, full as it was of conspiracy ingredients so typical for American cinema of the 1970s, seen, for instance, in films like *The Parallax View* (Alan J Pakula, 1974), *Three Days of the Condor* (Sydney Pollack, 1975) and *All the President's Men* (Alan J. Pakula, 1976) (see Ebert, March 16, 1979 and Canby, March 16, 1979). However, 12 days later the content of *The China Syndrome* suddenly became authentic and the film's involuntary 'political' message doubtlessly contributed to the negative psychological effect regarding nuclear power that swept throughout populations of the Western world. The film went to become one of the top grossers that year, taking a respectable $51 million at the US box office alone (Box Office Mojo, October 4, 2011). Eventually it was rewarded with three of the five major Academy Award nominations by the industry, a feat which set it apart from another related 1970s genre, the spectacular disaster film comprised of huge commercial successes such as *The Poseidon Adventure* (Roland Neame, 1972), *The Towering Inferno* (John Guillermin, Irwin Allen, 1974) and *Earthquake* (Mark Robson, 1974)—all nominated for several Oscars, but without fail for technical achievements.[1]

Based on the combination of its commercial success and the prestigious Oscar nominations, one could argue that *The China Syndrome* singlehandedly transformed, albeit inadvertently, the ecologically themed entertainment film from a genre of pure commercial thrills into a platform for environmental-political messages. Then again, all films are in some way driven by an agenda and contain opinionated messages, although mainstream Hollywood products have a tendency to downplay controversial political subjects. The Academy of Motions Pictures Arts and Sciences is even more conservative in its choices, not least due to the fact that there is at least one generation between Academy members and active film-makers, and two generations between Academy members and the average filmgoer, or as Oscar historian Emanuel Levy puts it: 'The Academy's preferences are always for safe, mainstream, noncontroversial films imbued with widely acceptable values' (2003: 48, 149, 364). Hence, an Academy Award nomination indicates that a, perhaps formerly, controversial topic has been elevated to the category of social importance and value. The quintessential example here is Hollywood's problematic relation to depictions of the Holocaust, a topic that was shunned in mainstream cinema for decades until the NBC miniseries *Holocaust* (Marvin J. Chomsky, 1978) finally broke the taboo, leading to the release of a wave of Holocaust films and television programs (Gustafsson 2010: 45–49). This becomes evident when studying the chronicles of Academy Award nominations where the Holocaust theme emerges from nearly non-existent to almost a mandatory facet of prestige productions, starting with documentaries like *Der Gelbe Stern* (Dieter Hildebrandt, 1981) and *Genocide* (Arnold Schwartzman, 1982), and then continuing with dramas like *Sophie's Choice* (Alan J. Pakula, 1982).

However, the acceptance of a film's message as an 'Oscar theme' does not automatically indicate that that theme becomes a commercially successful concept, even if it certainly helps—only that the subject matter can now be handled within the mainstream film production. In other words, the Academy Awards gives the 'message' or theme a certain amount of significance and credibility, whereas the process of mainstreaming enables that the message can be spread to a much wider audience than before. An Oscar nomination therefore serves like a measurement, both for the film industry's internal appraisal and as a pointer for the audience when it comes to current and noteworthy topics (Levy 2003: 372).

From this perspective *The China Syndrome* provides the origin for a new batch of environmentally aware films that, by now, have spread their 'message' through all sorts of genres, from the solemn drama and the factual documentary, to the self-conscious horror film and the big budget disaster movie. Depending on the sociopolitical context, these ecofilms have been subjected to both stimulation and disruptive conditions that alters the uses of styles, stories, and the different 'ecomessages', which in turn can be more or less important to an audience. Following on from this, the aim of this chapter is twofold. First, to present a historical overview of the mainstream ecofilm genre based on a survey of winners and nominated films for all categories of Academy Awards in the years 1979–2011.[2] Second, to analyse the development of ecocinema by discussing, not only the frequency but also how different themes, narrations, styles and notions of gender have been employed to evoke ecocritical consciousness, at the same time as these mainstream films had to entertain their audience.

The Origin of the Ecofilm: *The China Syndrome*

Although *The China Syndrome* had a great impact, this should not be misunderstood as if this individual film was the first ecofilm, that is, an entertainment film that in some way uses ecological awareness in its storytelling. During the 1970s a number of ecofilms appeared, and according to Robin Murray and Joseph Heumann, these were produced as a direct result of the impact of the first Earth Day in 1970 and the creation of US Environmental Protection Agency (EPA) the same year. The authors discuss films like *The Omega Man* (Boris Sagal, 1971), *Silent Running* (Douglas Trumbull, 1972) and *Soylent Green* (Richard Fleischer, 1973) as signifying models for a new masculine ecohero, based on ecological memories and nostalgia for the environment (Murray and Heumann 2009: 91–107). Nonetheless, these films were above all categorized as science-fiction films by contemporary reviewers and the ecological theme, if commented, was set aside as a trifling backdrop to the main story (See, for example, Ebert: March 10, 1972 and April 27, 1973, respectively). Needless to say, none of these films got recognition from the Academy regardless of their ecological theme.

Pat Brereton, on the other hand, presents a very wide and inclusive definition of an ecofilm. He locates the origin of the modern ecoaware film to the ecological fears, mostly based on atomic energy gone astray, that were exploited in science-fiction films of the 1950s such as *Them!* (Gordon Douglas, 1954) and *The Incredible Shrinking Man* (Jack Arnold, 1957). He also includes a wide range of films that, once again, could be labelled as science-fiction films, *Men in Black* (Barry Sonnenfeld, 1997), or disaster movies, *Titanic* (James Cameron, 1997), based on the assumption that even the exposure of nature in excessive imagery is enough to analyse it as an ecoaware film that actually educates its audience (Brereton 2005: 11–50.)

While ecophilosopher Bryan Norton states that green politics have to educate the public 'to see problems from a synoptic, contextual perspective' (1991: xi), it could be debated whether Hollywood has had, from a historical perspective, the ability or the *raison d'être* to educate its audience in that specific way. I do not disagree with the notion that the cinematic medium, especially the Hollywood mainstream film with its global impact, can teach different things to its audience. The effect of repetition and normalization is formidable in consolidating and making things familiar by means of moving images and sound. Still, if the audience do not know of, or care about, the tentatively and supposedly inherent 'message', then it is also doubtful if the same audience actually learns anything about ecology. Besides this, basic genre theory tells us that film-makers and filmgoers must, as a minimum, agree on a film's particular genre for it to work as such (Altman 2004: 144–65).

In other words, there is a great difference between films that are anachronistically interpreted as ecofilms, and films that consciously make an effort, either to say something serious about the environment or simply just to use a contemporary subject for commercial reasons. Thus, when Brereton exclusively performs textual analyses and then suggest that the American science-fiction films of the 1950s were imbued with 'critical ecological discourses'

(2005: 142), this should be contextualized by the fact that these B-films were aimed at a teen-oriented audience, mainly to be consumed at drive-in cinemas throughout the States. Consequently, the ecoawareness of these films was, at best, limited. A similar critique has also been raised against the ideological 'communist' readings of the same science-fiction films of 1950s (Keane 2001: 12–13). In this context it should be mentioned that among all these low budget science-fiction films of the 1950s, only two were acknowledged by the Academy, namely *The War of the Worlds* (Byron Haskin, 1953) and *Them!* which both were nominated for Best Special Effects.[3]

At the same time Brereton asserts that 'the very idea of being "green" only came into popular consciousness in the late 1970s' (2005: 12), something which corresponds well with the establishment of the anti-nuclear movement in the aftermath of the Three Mile Island accident and popular audiovisual expressions such as *The China Syndrome*. As a drama thriller *The China Syndrome* fits favourably into the formula of classical Hollywood narration with its focus on psychological motivation and character development, rather than on the excessive use of *mise-en-scène* to create an attractive spectacle. The story development is linear and follows three main characters, television journalist Kimberly Wells (Jane Fonda), cameraman Richard Adams (Michael Douglas), and nuclear plant supervisor Jack Godell (Jack Lemmon), as they discover and disclose an accident at the fictitious Ventura nuclear power plant in southern California. Following the conspiracy recipe of the 1970s, the accident turns out to be the result of corporate greed and the main antagonists consist of cowardly news directors and, above all, the evil nuclear corporation in the form of an anonymous security force that harasses Godell, and almost kills Adams' business partner Hector in a car accident as he is about to deliver evidence of forged x-ray films to the Nuclear Regulatory Commission (NRC).

Thriller elements such as suspense and excitement are blended with ingredients of melodrama, that is, a plot that deals with human emotions and personal crisis, though *The China Syndrome* avoids the mandatory romance plot which more often than not is intertwined with the main storyline in classical Hollywood cinema. This omission leaves room for those educational scenarios which Norton asks for, teaching the audience about the dangers of nuclear power via the dramaturgically biased explanations given by the audiovisual account of the first incident; the clarification of the term 'the china syndrome' at the convention of nuclear sciences; and by the statements on nuclear power made at a NRC hearing in the film. Even the original taglines for the picture are in a sense educational, as they pose variation on the statement: 'The China Syndrome. Today, only a handful of people know what it means... Soon you will know'.

However, the cause and effect of the story leaves the audience in ambiguity because the film uses corporate greed as the mystifying cause instead of arguments for or against the use of nuclear power in general. Or as the reviewer in *The New York Times* writes: '"The China Syndrome" is less about laws of physics than about public and private ethics' (Canby: March 16, 1979). The thriller elements are also dependent on the acting and character development rather than on the tickling excitement of pyrotechnics in the *mise-en-scène*, even if that occurs on a minor scale by the end of the film. In order to create suspense the SCRAM (the plant's shutdown system) causes significant damage to the cooling system which collapse in a spectacular scene, at the same time as Godell is shot by a SWAT team as he tries to prevent the catastrophe. These elements from the disaster movie genre of the 1970s are well executed and were also rewarded by the Academy with a nomination for Best Art Direction and Set Decoration. Nevertheless, the main focus was on the acting and both Fonda and Lemmon were nominated for Best Actress and Actor, respectively, in a Leading Role, accompanied by a third major nomination for Best Screenplay, directly written for the screen.

In the end, *The China Syndrome* is closely related to the American conspiracy thriller of the 1970s and would probably have continued to be grouped together with films like *All the President's Men* and *Capricorn One* (Peter Hyams, 1978), were it not for the nuclear accident at Harrisburg which, in an instant, changed the reception of the film, transforming it into a serious ecofilm with elements of the thriller, rather than vice versa. The accident probably contributed to the Academy's decision to elevate *The China Syndrome* from a mere dime a dozen thriller to Oscar prestige. In any case, this laid the basis for how an ecofilm should look like, as we shall see, both for the Academy's voting members and for the broad cinema-going audience.

Oscar Nominations and the Ecofilm

After the success of *The China Syndrome*, one would think that a number of easily identifiable ecofilms would have been rushed into production as a way for Hollywood to cash in on the latest craze. Yet, Hollywood does not always follow the yellow brick road and the ecoaware film does, in fact, continue to live on the border between different subgenres. According to Emanuel Levy, an Oscar Theme lives in cycles of 3 to 5 years, showing up, making an impression, and then disappearing for a number of years until the theme returns in a slightly different form than previously (2003: 149).

The survey of the Academy Award's nominations for the years 1979 to 2011 clearly shows sign of this cyclical behaviour, with evidence of four cycles of ecofilms. The first one begins modest with a technical nomination for the disaster movie *The Swarm* (Irwin Allen 1978) and then continues up until 1985, including twelve films and 32 nominations, of which the most noticeable are *The China Syndrome* (four nominations), *Silkwood* (Mike Nichols, 1983, five nominations), *The River* (Mark Rydell, 1984, four nominations) and *Places in the Heart* (Robert Benton, 1984, seven nominations). The second one goes on from 1988 to 1991 and includes three films and seven nominations, of which *Gorillas in the Mist: The Story of Diane Fossey* (Michael Apted, 1988) is the most obvious with five nominations. Thereafter, the ecofilm all but disappears from the Academy Awards until it reappears for a third cycle (1996–2001, four films and ten nominations) with technical nominations for *Waterworld* (Kevin Reynolds, Kevin Costner, 1995) and *Twister* (Jan De Bont, 1996), and culminates with the clearly ecologically themed *Erin Brockovich* (Steven Soderbergh, 2000, five nominations). The fourth and final cycle stretches from 2006 to 2011, and is possibly still ongoing. This period includes ten films and 27 nominations with such ecoaware films as *An Inconvenient Truth* (Davis Guggenheim, 2006), *Michael Clayton* (Tony Gilroy, 2007) and *The Cove* (Louie Psihoyos, 2009).[4]

All in all, the Academy awarded 29 films with a total of 81 nominations between 1979 and 2011.[5] These statistics illustrate the fact that the concept of ecology has worked quite poorly as an Oscar Theme over the last 30 years. The entire sum of nominations between 1979 and 2011 amounts to an approximately figure of 3200, of which the ecofilm only received 2.5 per cent. This can be compared to the nominations for Dramas (48.6 per cent), Comedies (18.2) and Action-Adventures (6.1) in the years 1927–2001 (Levy 2003: 385). There have, of course, been a number of ecofilms that never made the Oscar cut, ranging from small films like *The March* (David Wheatley, 1990) to big budget film like *The Day After Tomorrow* (Roland Emmerich, 2004). Nevertheless, the low percentage reflects, in a way, a somewhat indifferent attitude towards the environment in Western societies, and this in spite of all the alarm reports on pollution and the overexploitation of nature's resources that have figured in the news since 1979. Why is this? A first conclusion would be that there is no correlation between what goes on in the film industry and the outside world. Then again, film is a popular art form and there is, of course, a correlation as to how historical, political and economic factors influence Hollywood and the Academy. The question is in what way the ecofilm developed throughout these four cycles, and how this change, when it came to portraying ecoawareness, can be connected to a sociopolitical context.

Cycle 1: The Emergence of the Ecoheroine

The first cycle of Oscar nominated ecofilms were heavily influenced by three features: the continued close connection in style to the 1970s disaster movie, the recurrent theme of nuclear energy gone wrong, and the fact that the portrayal of the main characters was often based on class issues with a focus on ordinary people of both sexes rather than on extraordinary heroes or, even less frequently, heroines (Toplin 2002: 30–32). The last disaster movies of the 1970s era, *Meteor* (Ronald Neame, 1979) and *When Time Ran Out* (James Goldstone, 1980), were also among the films that the Academy nominated for technical awards in this cycle. However, a 1970s feature that became virtually extinct in the first cycle was the conspiracy component. Not even in a film like *Silkwood*, with its connection to real-life labour union activist Karen Silkwood and her mysterious death, was this trait highlighted. This does not mean that the concept of an abstract antagonist, as seen in *The China Syndrome* or *All the President's Men*, was discarded, just that the antagonist became more bodily present. Corporate greed, however, retained its position as the prime kind of evil in many of these films and there is a reason for this.

In film studies, the 1980s is usually associated with reactionary Reaganism and the emergence of the male centred action film with Sylvester Stallone or Arnold Schwarzenegger as masculine superheroes in films such as *First Blood* (Ted Kotcheff, 1982) and *Commando* (Mark L. Lester, 1985). Albeit commercially successful, these films were not in any way acknowledged by the Academy. On the contrary, the appearance of the ecofilm coincided with a batch of class-aware films, such as *Norma Rea* (Martin Ritt, 1979), *Reds* (Warren Beatty, 1981) and *The River*, that all did well in the nominations for Oscars. In fact, the ecoaware film often overlapped the class-aware film, especially in films like *Silkwood* and *The River*, which reveals that the conservative Academy had a distinct leftwing outlook at the time. This, in turn, also marks the establishment of yet another recurrent characteristic of the ecofilm that surfaces at this time, namely the female ecoheroine. In contrast to the tragic ecohero of the 1970s and the pumped up macho man of the 1980s, the ecoheroine got the major nomination for Best Actress in a Leading role on several occasions: *The China Syndrome* (Jane Fonda), *Silkwood* (Meryl Streep), *Testament* (Lynne Littman, 1983, Jane Alexander), *The River* (Sissy Spacek), *Places in the Heart* (Sally Field) and *Country* (Richard Pearce, 1984, Jessica Lange).

I believe that the ecoheroine is essential for understanding the development of the American ecofilm. The disaster movie of the 1970s had been filled with ordinary but resourceful men often typecast with actors such as Charlton Heston and George Kennedy. Women, on the other hand, had been reduced to helpless objects in this genre, either subjected to the gaze of men or ridiculed as hysterics, which can be observed in a film like *Earthquake*, with Genevieve Bujold's beautiful single mother and Ava Gardner's possessive alcoholic housewife embodying these characteristics respectively. One important reason for this old-fashioned outlook on gender was the fact that the disaster movie was extremely expensive to produce. A consequence of this was more conservative portrayals of gender than, for example, in melodramatic features based on acting rather than pricey explosions (Keane 2001: 15).

When the dominant image of masculinity moved on to the action-adventure this left room for larger and more independent female roles, not least in the newly invented ecogenre. However, none of these ecoheroines were tough in a physical, bodily way—as Ripley (Sigourney Weaver) in the *Alien* franchise—but rather, their strength was of the mental variety, persevering against different hardships and setbacks. Although many of them were portrayed traditionally as caring mothers, wives or girlfriends, not all of them had a nuclear family to take care of. In this way the ecofilm came to be connoted as a feminine project and the portrayal of educational ecoawareness was, for a number of years, mainly performed in the genre of melodrama.

In Hollywood, melodrama and the woman's film have had a long history as a fruitful site for feminist criticism. As Yvonne Tasker puts it: 'Melodramatic narratives and the woman's film can be understood as both foregrounding and testing the contradictions between desire and duty in mainstream definitions of femininity' (1998: 141). Pat Brereton borrows the term *Ecofeminism* from French writer Francoise d'Eaubonne in order to explain what he sees as an explicit feminist discourse in environmentalism. According to Brereton, feminist discourse is inextricably linked to green politics because it holds an 'inherent critique of masculinity and its "values", as well as a critique of rationality and the "overvaluation of reason"' (2005: 31–32).

The division between the disaster movie as a masculine genre and the melodramatic ecofilm as a female genre becomes even clearer when looking at the films and their reception. *Silkwood*, like its predecessor *The China Syndrome*, has a woman at its centre, but unlike *The China Syndrome*, the story in *Silkwood* moved into the private sphere with a focus on different relationships at home and at work which, in this instance, happened to be the Kerr-McGee nuclear plant in Oklahoma. *The China Syndrome*'s thriller elements, on the other hand, did not engage in this type of private relationships, bending it into a more masculine film. One example is that the character Karen Silkwood (Meryl Streep) was portrayed in a more bodily manner than Kimberly Wells (Jane Fonda) in *The China Syndrome*, both when it came to sexual relations but more importantly, when it comes to the matter of being personally affected by the 'evil' of nuclear power, since Karen was both contaminated and even died in a car accident at the end, possibly at the hands of the evil corporation.[6] As Linda Williams has argued, melodrama or the 'women's film', is one of the body genres that are designed to bring out sympathy and strong emotions when seeing the misfortunes of onscreen characters (2009: 602–16). What is more, when these strong emotions are intricately associated with nuclear contamination, or other ecological issues, the genre becomes a vehicle for the (of course biased) education about nuclear energy.

This ambivalence between information and emotion can also be observed in the contemporary reception of *Silkwood* where it was described as a 'highly emotional melodrama' where the 'muddle of fact, fiction and speculation almost, though not quite, denies the artistry of all that's gone before' (Canby: December 14, 1983). Another reviewer believed that the film was 'an angry political expose' before seeing it but corrected himself and wrote that the film 'isn't about plutonium, it's about the American working class' (Ebert: December 14, 1983). Then again, *Silkwood* was also described as 'the first [...] popular movie [where] America's petrochemical-nuclear landscape [had] been dramatized [...] with such anger and compassion' (Canby: December 14, 1983). In fact, the two sides of the coin become hard to separate when melodramatic components drive the plot and factual ecoawareness is mixed up with the display of strong emotions. And it was exactly this combination that gave a film like *Silkwood* its weight in the rising ecological debate.

While Silkwood can hardly be described as visually exciting, it was nevertheless elevated to the category of a socially important and valuable film by the Academy and received five nominations, including three of the major ones for Best Actress in a Leading Role, Best Director and Best Original Screenplay. Similarly, in a film that took the threat of nuclear power to the extreme, Jane Alexander was nominated for Best Actress in a Leading Role for *Testament*, about which Roger Ebert wrote:

She stands by her children, watches as they grow in response to the challenge, cherishes them as she sees all her dreams for them disappear. It is a great performance, the heart of the film. In fact, Alexander's performance makes the film possible to watch without unbearable heartbreak, because she is brave and decent in the face of horror.

(Ebert: November 4, 1983)

At the Academy Awards the year after, the effect of the ecoheroine became prominent as the character swept the awards with three nominations for Best Actress in a Leading Role for Sissy Spacek, Jessica Lange and Sally Field in *The River*, *Country* and *Places in the Heart* respectively, eventually landing the Oscar statuette in the hands of Field for the second time around. All together these three films earned 12 nominations and two wins. These three films went under the name 'country movies' in 1984 since all of them dealt with the life of American family farmers in the face of an unforgiving nature and financial problems. This mirrored the contemporary credit crunch brought on by low-price farm loans and falling market prices, even if *Places in the Heart* took place during the depression in the 1930s.

The connection between melodrama and the ecoheroine is reinforced via these films, where the woman's strength holds things together in the fight against nature, stubborn and stupid husbands or the equivalent, and not least the evil corporation in the form of the bank. Their educational value regarding ecoawareness of nature is mainly apparent through the use of magnificent imagery of fields, crops, livestock and the weather—most notably in *The River* where the cinematography was nominated for an Oscar and one reviewer duly commented: 'Vilmos Zsigmond has photographed things so dewily that even the disastrous floods look more pretty than terrifying. Life is hard and life is tough but [...] it's also photogenic' (Canby: December 19, 1984). These would therefore, according to Brereton, be examples of the 'therapeutic romantic power of nature' that helped viewers deal with the everyday distress of modern living (2005: 17, 21).

But contrary to Brereton's believes that this form of ecounderstanding can be dangerous as it imposes a false consciousness on the audience about the environment; I believe that these romantic representations of nature have more purpose than simply being therapeutic. First, nature is portrayed as unruly in spite of, or perhaps because of, the fact that farming essentially is man's (attempt at) advanced control of nature. The melodramatic down-to-earth portrayal of nature therefore serves as an education, in this case pertaining to agriculture, which enlightens the audience about the value of the environment. Second, in all of these films the farmers, that is, the women act as symbolic guardians between nature and the big business corporation. In hindsight this female guardianship becomes more than just symbolic since this period, i.e. around 1980, marks the final transition from small family farms to the establishment of giant American food manufactures. This transition resulted in a bigger gap between nature and the responsibility of human beings given that corporations often are said to have neither consciences nor respect for nature (Bakan 2005). This in turn makes women's struggle true, believable, and certainly melodramatic and sentimental— hence the objections that romantic representations of nature can be misleading. One example of this unsympathetic misinterpretation can be observed in relation to *The River*. Contemporary reviewers unanimously pointed out that this film mainly ran on female sentimentality. At the same time, they were critical of the fact that Scott Glenn's character Joe Wade became the villain just for the reason that he introduced ways to deal with the effects of nature that sounded like common sense in their ears (Ebert: January 11, 1985 and Canby: December 19, 1984), while in fact these ideas represented the first step towards big business faming.

Gender is by definition a relational concept, and the emergence of the ecoheroine was no exception (Appelby, Hunt and Jacob 1997: 209–18). As successors to the tragic ecohero, the ecoheroines also acted as pioneers and champions of the environment, but unlike the tragic ecoheroes they did not act as prophets of doom, nor did they rely on a nostalgic sense of nature (Murray and Heumann 2009: 94–96). Then again, the ecoheroine can be connected to the durable Western belief that women are closer to 'nature' than men, whom in this binary thinking are thought to be closer to 'culture'. Accordingly, nature has throughout history been treated as inferior to culture, and humans have been understood as being separate and superior to the natural environment (The Green Fuse 2011).

However, in the context of these green films, the ecoheroine—in reviews described in explicitly masculine terms as 'staunch and heroic' (Canby: December 19, 1984) and 'brave, stubborn' (Ebert: January 11, 1985)—overtake the farming men at their own game for two reasons. First because of the more or less complete male failure where masculine icons such as Mel Gibson (*The River*) and Sam Shepard (*Country*) are portrayed as weak, or simply as inferior due to race (Danny Glover in *Places in the Heart*) or physical disability (John Malkovich in *Places in the Heart*). Typically, Malkovich received the only male acting nomination as a blind boarder that year. The second reason can be linked to their critique of patriarchal society, built on sexism, racism, class exploitation and environmental destruction, in view of the fact that big business thinking per definition was exposed as immoral in these films. The only male character that was not depicted as a weakling (Scott Glenn in *The River*) was also dismissed in the cinematic context as an evildoer just because he proposed that nature could be altered and controlled.

Female characters had strong roles in these ecofilms that, unfortunately, were scarce and far apart. Nonetheless, partly for the reason that the Academy helped to familiarize these ecofilms for a larger audience, the establishment of the ecoheroine was firmly in the hands of this sub-genre of ecofilms. This was, as well, similar to the development of female leads in another body genre during the same period. As Carol Clover has suggested in relation to the horror film, the concept of gender was more equally distributed in gendered roles in the low budget horror genre than mainstream cinema. Furthermore, according to Clover, these progressive tendencies moved into the mainstream cinema, for example in films such as *Thelma and Louise* (Ridley Scott, 1991) and *Silence of the Lambs* (Jonathan Demme, 1991) (1992: 3–20). Although not explicitly connected to the development of the horror film, the same discourse can nevertheless be observed in the ecofilm during the early to mid-1980s. The educational value of these ecofilms was thus linked to a feminist viewpoint, both in relation to environmentalism and when it came to the equitable representation of female characters.

Cycles 2 and 3: The Ecological Void

Contrary to common belief, the documentary genre was not overrun by ecological themes, at least not during the first three cycles where the absence of nominations for Best Documentary Feature is particularly peculiar. Only three nominations were handed out by the Academy to documentaries that could be labelled as ecoaware between 1979 and 2005. The first one was for *In Our Water* (1982) which dealt with a family from South Brunswick, NJ, whose water had been contaminated by lethal poison from a nearby landfill. This was a low budget production from onetime producer/director Meg Switzgable. The other two nominations came for *Radio Bikini* (Robert Stone, 1988) and *Building Bombs* (Mark Mori, Susan Robinson, 1991), the latter narrated by actress/activist Jane Alexander.

Both these films dealt with the nuclear weapons build-up, but not as a danger of World War III. Instead these documentaries focused on the American nuclear industry as a threat to the local environment and population in the Bikini Atoll, The Pacific Ocean, and the state of South Carolina, respectively. The showcasing of 50 years of bureaucratic negligence, corporate greed and sheer stupidity would have developed into a persuasive ecological critique, similar to the criticism raised in *The China Syndrome*, were it not for the fact that the US government and the American nuclear industry appeared as insensitive psychopaths. Whereas conspiracy ingredients could be handled as 'entertainment' in political thrillers, this sort of argumentation was not perceived as well with the documentary genre due to its association with truthful representation. *Building Bombs* was, for example, banned from airing on the PBS-network due to the fact that the film was so openly politically biased (IMDb: November 1, 2011 and PBS 2011).[7] In other words, the ecotheme in these films was not deemed to be significant enough in order to warrant their partisan takes on the endangered environment. In addition, both films adopted a historical perspective that basically undermined their contemporary validity, given that the ecological problem was mainly relegated to the past. A parallel case in point would, for instance, be Michael Moore's two documentaries *Bowling for Columbine* (2002) and *Fahrenheit 9/11* (2004). These films were heavily contested as biased productions at the same time as they were hailed as imperative vehicles for current political issues. *Bowling for Columbine* was even rewarded by the Academy for Best Documentary Feature in 2003 despite various allegations that Moore's film indeed was not a proper documentary.

The suggestion that the documentary genre could not face environmental problems head-on is, certainly, a somewhat one-sided conclusion since the Oscars are not deterministic when it comes to the overall film production, especially pertaining to American independent documentary film-making. However, most of the documentaries nominated between 1979 and 2011 were independent productions and this, nevertheless, gives a strong indication of the ecotheme's weak position as an essential topic of contemporary relevance.

The absence of ecoaware films remains the most obvious sign of this extended period that included cycle 2 and 3 and which ran from 1986 until 2005. Even the feature film, which began with a row of nominations during the first cycle, no longer seemed to embrace the ecological theme. In fact, only four ecofilms were nominated during this 20-year period, of which two dominated their respective nomination years. In 1989, *Gorillas in the Mist* got five nominations including Best Actress in a Leading role for Sigourney Weaver, and in 2001 *Erin Brockovich* won the Best Actress statuette for Julia Roberts. Sure enough, these films had an ecoheroine thereby extending the tradition with strong female leads in ecofilms. The other two films were the disaster movie *Twister*, again with a strong female lead in Helen Hunt, and *A Civil Action* (Steven Zaillian, 1998), like *Erin Brockovich* a melodramatic court room drama about environmental pollution and corporate greed with an upbeat ending, however, this time with a male ecolawyer in the guise of John Travolta.

While *Erin Brockovich* and *A Civil Action* resemble each other, much in the same way as *The China Syndrome* and *Silkwood*, and later the country-trilogy, resembled each other in previous cycles, *Gorillas in the Mist* stands out because of its explicit display of nature. One would think that nature, whether portrayed romantically or not, should be a crucial part of the ecofilm but that has actually not been the case. However, much of *Gorillas in the Mist* was photographed on location in the mountains of the Ruhengeri region in Rwanda, which gave the audience a spectacular view of wildlife nature—as compared, for example, to the restrained nature in *Places in the Heart* or the non-existent nature in *Silkwood*. As a consequence, the *mise-en-scène* in itself—most notably via the display of mountain gorillas—took on a significant role, almost equal to that of the actors. Thereby the film also works as an example of a didactic ecofilm, but *nota bene*, in the 'primitive' way that appealed more to emotions than to reason.

Regarding *Gorillas in the Mist*, Brereton states that the story of Diane Fossey 'provides a suitable platform for ecological exposition', but nevertheless that the high-cultural intelligentsia 'often fails to recognize the potential of popular culture discourse which requires subtlety and sophistication in its delivery' (2005: 59). Even so, the contemporary reception showed that the therapeutic romantic power of nature actually did work on the 'intelligentsia' as many expressed strong emotional experiences when it came to the numerous scenes between Sigourney Weaver and mountain gorillas. Janet Maslin in *The New York Times* wrote that these scenes 'have an irresistible appeal' and that the 'glimpses of the mountainous Rwanda [...] are indeed breathtaking' (September 23, 1988), while Hal Hinson, after trashing the film for its bad script, still admitted that 'whenever the cameras turn on the gorillas [...] you feel you're witnessing something truly great' (September 23, 1988). The usually harsh film critic Roger Ebert confessed that '[t]here were moments when I felt a touch of awe. Those moments, which are genuine, make the movie worth seeing' (September 23, 1988).

A comparison with another heavily awarded film, *Dances with Wolves* (Kevin Costner, 1990, 12 nomination and 7 Oscars), that is not included as an ecofilm in this chapter, reveals that the landscape, although vast and beautiful, does not assume the same implication of ecoawareness as it does in *Gorillas in the Mist*. Perhaps this is due to the historicizing of the landscape in *Dances with Wolves*, where the astonishing features of the *mise-en-scène* are simply turned into ancient history, regardless of the fact that the film was shot on location in modern-day South Dakota. In addition to this, several film critics pointed to the fact that the film was produced with an air of 'all-American boyishness' (Hinson: November 9, 1990) and that 'Costner [...] creates a vision so childlike, so willfully romantic, it's hard to put up a fight' (Howe: November 9, 1990).

A conclusion to be drawn from this is that fact and fiction work as dividing factors in relation to ecoawareness. That is, although *Dances with Wolves* certainly created awe among its audiences, this did not transform into ecoawareness as it did with *Gorillas in the Mist*. The historical setting and the boyish qualities basically worked against the 'inherent' educational value of the film, albeit not as a history lesson but more as an ecological concern. One could argue that the exotic setting of Rwanda in *Gorillas in the Mist* had become an ecological surface *per se* due to the numerous wildlife documentaries on TV, whereas the landscape in *Dancing with Wolves* had been normalized as a result of the Western genre (Chris 2006). A look at all ecofilms nominated for Oscars between 1979 and 2011 also reveals that not a single film can be labelled as historically fantastic or comic, with exceptions of the two animated films *Finding Nemo* (Andrew Stanton, Lee Unkrich, 2003) and *Ice Age* (Chris Wedge, 2002). Instead they are contemporary, melodramatic and solemn. In other words, ecofilm is no laughing matter.

Sigourney Weaver's towering and physical performance as Diane Fossey does not only extend the tradition of strong female leads in ecofilms, but it was also a reminder of her role in the *Alien* franchise. Consciously or not the portrayal of Fossey, like that of Ripley, has a duplex quality over it since Weaver demonstrates nourishing, protecting, and violent attributes *vis-a-vis* the environment. This heroism differs from earlier, and later, portrayals of ecoheroines where sheer physicality was toned down in an otherwise melodramatic setting. What is more, this seemed to confuse contemporary reviewers seeing that they agreed on the unbalanced screenplay that left too many loose ends unsolved that were partly linked to the female encoded violence and partly to the mysterious death of Fossey in the film (see Maslin and Hinson: both September 23, 1988). Both these circumstances can be linked to the question of the environment as well. First, was it justified to use violence to protect nature, i.e. the mountain gorillas? Second, by not answering the question of whom murdered Fossey the film does not just fail to give the audience closure on a emotional level but also on an ecological-educational level since the question of whom murdered Fossey irrefutable was linked to the question of ecology in the sense that somebody—trappers, pygmies, the Rwandan government—gained pecuniary from the murder.

If *Gorillas in the Mist* differed from how the ecofilm, as appreciated by the Academy, was set up, *Erin Brockovich* and *A Civil Action* continued and consolidated the melodramatic tradition with focus on the individual and the diegetic romantic use of nature as something that happened outside of the courtroom. Tellingly, *Erin Brockovich* had turned into a 'China Syndrome Lite' (Hays: March 10, 2000) and in reviews, comparisons with *Silkwood* and Meryl Streep were plentiful and not to the advantage of Julia Roberts. In fact, reviews focused almost solely on Roberts' performance, calling attention to her 'unwise wardrobe' that positioned the character 'somewhere between a caricature and distraction' (Ebert: March 17, 2000). She was called a '[m]iniskirted hooker' (Ebert: March 17, 2000) and the film felt like 'Pretty Woman meets Flashdance' (Hays: March 17, 2000). In *Rolling Stone Magazine* this was seen as something positive as *Erin Brockovich* was perceived as 'outrageously, even shamelessly, entertaining' and Roberts as 'tough, tender, sexy and brashly funny' with a 'wardrobe of a hooker' (Travers: March 17, 2000). And in *San Francisco Examiner* the same clothes were said to come from *Pretty Woman* (Garry Marshall, 1990), something that apparently 'compromised' the message of the film (Morris: March 17, 2000). Or as Roger Ebert wrote: 'If the medium is the message, the message in this movie is sex' (March 17, 2000).

Thus, *Erin Brockovich*'s ecological theme seemed to be upstaged by film star Julia Roberts in her Oscar vehicle because the film was not taken as seriously as an ecofilm should. Furthermore, the film was slandered as 'a standard-issue do-gooder melodrama', a 'Hollywood soap opera, a smooth second-hand amalgam of "Norma Rae" and "Silkwood"' (Scott: March 17, 2000) and 'Erin Brockovich, in life and on the screen, can't sell candy-assed platitudes about the triumph of the human spirit [...] because, mercifully, she sucks at it' (Travers: March 17, 2000). This gendered bashing, realized by male reviewers that essentially, almost exclusively, chose to talk about what clothes and behaviour Roberts (and indirectly Brockovich) wore and demonstrated, belittles and obscures the ecological theme of *Erin Brockovich*. Not surprisingly the critique in the reviews was also linked to the film's melodramatic implementation, and thereby to this genre's supposedly female qualities, that was disparaged in a routine manner.

In feminist film criticism the film melodrama has been characterized as subversive, as an expression of bourgeois realism, as popular entertainment and, perhaps above all, as a markedly female genre—both in terms of the subjects depicted and their intended audience—which provides a partial explanation for its low social status (Altman 2004: 5; Soila 1991: 35–44).[8] However, as Rick Altman has pointed out, the designation 'woman's film' had never been associated with the film melodrama prior to 1970. The melodrama was renegotiated when feminist film scholars wished to rehabilitate female activities on film, transferring them from family melodramas to a women's genre (2004: 1–77). This gendered coding has meant that the film melodrama has almost exclusively been studied from a feminist perspective, focusing on female identity and 'female' activities—and thus the genre has come to be perceived as a 'women's genre'.[9] Unfortunately, this has led to an underestimation of the historical qualities and possibilities of the melodrama, which ultimately cements the gender dichotomy on the basis of predetermined patterns of power.[10]

As a consequence, when Julia Roberts as Erin interviews people whom have been exposed to poisonous chromium—causing death and cancer—and tries to convince them to join the class-action lawsuit, she is, via the predisposed melodramatic concept, already reduced to a feminine stand-in (in a miniskirt) who feels their pain for the audience. The educational value of this totally bypassed the professional reviewers. Then again, when screening through the user reviews on IMDb the educational value becomes more apparent since nearly all posts focus on what she actually does, her activity, rather than on what she looks like—which is something that, as one user suitably put it, 'heightens its social aspect because we identify with the little man' (IMBd, November 6, 2011).

The gendered reception of *Erin Brockovich* becomes even more apparent when compared to the reception of *A Civil Action* with John Travolta as the ecohero. As in *Erin Brockovich*, *A Civil Action* does not use extensive visual illustrations of the environment, except during the film's prelude when Travolta, as the unscrupulous lawyer Jan Schlichmann, muddles his expensive shoes to discover 'the tannery that has been dumping evil-looking chemicals', as one reviewer so aptly put it (Maslin: December 25, 1998). Instead *A Civil Action* follows the melodramatic route to the letter, although the ending is more ambivalent than the feel good finale of *Erin Brockovich*. In spite of the apparent similarity in theme and execution, the reviewers did not talk about Travolta's wardrobe in the demeaning fashion as they had done with Roberts. Instead *A Civil Action* was seen as a 'character study' and a 'moral journey' (Lasalle, January 8, 1999), and parallels were even drawn to *Schindler's List* (Steven Spielberg, 1993) as 'Travolta, emerges as a Schindleresque figure—a large, imposing, slickly dressed opportunist who accidentally finds his soul' (LaSalle: January 8, 1999 and Maslin: December 25, 1998).[11]

The male ecohero was here, per definition, perceived as more serious than the female ecoheroine, even if the ecoheroine had been around since *The China Syndrome*. In fact, Julia Roberts' role as environmentalist Erin Brockovich was the latest one to be awarded by the Academy[12]; while the ecohero has had at least one strong follow-up in the seven-time nominated *Michael Clayton* with George Clooney as its ecohero/lawyer. However, between the end of cycle 3 in 2001 and the beginning of cycle 4 in 2006, the ongoing ratification of the Kyoto Protocol (initiated in 1997 with the aim at reducing greenhouse gas emissions) and the effects of climate change emerged into popular consciousness. Since this originally scientific discourse gradually grew more popularized and globalized, it was soon exploited by the film industry in a more straightforward way than before. In big budget films like *Waterworld* and *The Day After Tomorrow* the devastating consequences of climactic change were on display on a grand scale. This move from small scale ecomelodramas towards big budget disaster films once more shifted the individual focus from ecoheroines to ecoheroes—and once more the Academy nominations, as in the case of the disaster movies of the 1970s, was withheld with the exception of an occasional technical nomination. This development was, obviously, reflected in the mega success of *Avatar* (James Cameron, 2009) where the legendary ecoheroine Sigourney Weaver was sidelined for the benefit of male action-adventure fantasy in 3D. Tellingly, *Avatar's* 9 nominations were of technical nature, with the exception of the ones for Best Picture and Best Director, arguably given by the Academy on the account of its impressive box office record.

Cycle 4: The Ecodocumentary and the Male Ecohero

The fourth and final cycle is dominated by the ecodocumentary, starting with *Darwin's Nightmare* (2004) about the disturbing ecological effects of globalization as Austrian film-maker Hubert Sauper explores the fishing town of Mwanza, Tanzania, were the Nile Perch is caught, factory processed and then shipped out to Europe as a delicacy that no native African can afford. The EU was and still is, nonetheless, satisfied with this African success story and with the commercial fishing industry that exports 500 tons of fish fillet every day. From an ecological perspective Sauper's documentary produced a bleak and devastating effect—visually and content-wise—as the introduction of the Nile Perch to Lake Victoria in the 1950s has more or less destroyed the ecosystem, which ultimately has changed the living conditions of the local inhabitants for the worse, at least according to the film's storyline where globalization is presented as something inherently evil. Western reviewers were, accordingly, very receptive of the view of ecological globalization that was presented in *Darwin's Nightmare*, simply because it fitted into their (or our) world view. In *The New York Times* it was called 'a work of art' (Scott: August 3, 2005. See also Johnson: October 14, 2005) and in the European *Svenska Dagbladet* the reviewer wrote: 'It's horrible. But the worst thing is that it's true. If you should just watch one film about globalization, go see *Darwin's Nightmare*' (Gentele: November 17, 2005, my translation).

The exceptionally negative imagery can actually act as a sort of catharsis because it leaves the audience with the notion that somebody else has finally addressed a specific current problem. Sociologist Stanley Cohen has argued that mass media, especially television where documentaries find their main outlet, almost have a monopoly in creating cultural images of suffering. Cohen identifies several formal elements in this imagery, for example, that the use of negative imagery of suffering from developing countries—starving children, war victims, refugee camps—as the 'normal' state results in a cultural denial in the West that alienates rather than engages people in front of their TV-sets (2007: 169, 183–84). At the same time the cultural belief in visual images, and that they can have a visceral impact, continues to be strong:

> Sophisticated technology can spread images of live atrocities around the world in minutes. But self-evident truth will not be self-evidently accepted. However informative, reliable and convincing they are, accounts of atrocities and suffering do little to undermine overt forms of denial. Humanitarian organizations are living relics of Enlightenment faith in the power of knowledge: *if only people knew, they would act.*
>
> (Cohen 2007: 185, italics in original)

It should be noted that denial is a cognitive mechanism that necessitates what Cohen calls the denial paradox: that the denial in itself reveals that the person or country actually did know what they denied, otherwise it would not be necessary to deny anything. In view of this, Cohen dismisses traditional rationalizations of denial, such as the psychoanalytical defense mechanism of the unconscious and the thesis about compassion fatigue, due to information overload and desensitization. In its place he distinguishes several more precise basics of denial, among them *moral distance*, as repetition of images of suffering increases their remoteness (2007: 5–6, 187–91, 194).

Although recognized by the Academy, *Darwin's Nightmare* was, perhaps not unsurprisingly, neglected in favor of a more lighthearted ecodocumentary, namely *La Marche de l'empereur/*

March of the Penguins (Luc Jacquet, 2005), which in its American version with Morgan Freeman as narrator, at least, became more serious in its presentation compared to the childish European versions.[13] In any case, the gates were opened for ecodocumentaries and in 2007 the big winner was, of course, *An Inconvenient Truth*, former vice president Al Gore's epic slide show on global warming that developed into a phenomenon in itself with an impressive box office return of nearly $50 million worldwide, actually placing it as the 6th most seen documentary in US history (Box Office Mojo, November 7, 2011).

Although the film was subsequently subjected to negative critique, and Al Gore personally attacked by his political opponents—which among other things included a public congressional hearing on March 21, 2007 (YouTube 2011)—the initial reception was no less than astonishing. Robin Murray and Joseph Heumann claim that *An Inconvenient Truth* succeeded, 'not because of its predictions but because of the eco-memories it evokes' (2009: 196). They also claimed that this was an ideological project where the use of nostalgia created a false sense of belonging because the historic environment was already lost (2009: 201). That may be so but when reviewers were about to handle this film, they not just took the words and images in it *ad notam*; they gave a detailed account of the facts presented in the film and then added an urgent call to go see the film as if the world literally depended on it (see, for example, Ebert: June 2, 2006; LaSalle: June 2, 2006; Scott: May 24, 2006; Knipp: June 23, 2006; Thomson: June 2, 2006). Being a slide show, crosscut with personal reflections on Gore's life, the educational impact far exceeded most, if not all, previous ecofilms, whether it be melodramatic feature films or documentaries. In fact, the on the surface banal presentation worked primarily for three reasons: First because of the, as mentioned, heavy factual basis which was presented by Gore in a dry professor-like manner. Second, and equally important, because of the strong emotions that *An Inconvenient Truth* evoked. Reviewers used graphic words as 'shock' (LaSalle: June 2, 2006), 'disturbing' and 'gasps of horror' (Scott: May 24, 2006), 'alarming' (Macdonald: June 2, 2006) and 'scariest' (Biancolli: June 9, 2006) to describe the frightening experience of the film. However, the film would not have had the impact it did were it not for its emotional structure that gave room for hope in the end, as the problems with carbon dioxide emissions and global warming are presented as 'fixable' (Puig: May 23, 2006), even with exhortations of different climate-friendly actions during the end credits of the film, all while Melissa Etheridge's Oscar winning song 'I Need to Wake Up' plays.

It is hard to discuss *An Inconvenient Truth* without taking the man at its center, Al Gore, into account. Al Gore's persona is, in fact, the third reason which contributed to the success of this ecofilm. In spite of, or actually rather because of, Gore's media built image as a bore, the film comes off as a revenge of sorts, although it is not structured as a comeback story. Roger Ebert's recollection gives a hint of this, and he is not the only one: 'When I said I was going to a press screening of "An Inconvenient Truth", a friend said, "Al Gore talking about the environment! Bor…ing!"' (June 2, 2006). Due to audience knowledge of Gore's close loss in the 2000 presidential election—he even jokingly introduces himself as follows: 'I used to be the next president of the United States'—everything is compared with his earlier persona, and the outcome is a feeling of surprise of the fact that he is funny, laidback and real. Some call it 'anti-charisma' (Bradshaw: September 15, 2006), others see him as 'droll and relaxed' (Macdonald: June 2, 2006). Mick LaSalle in *San Francisco Chronicle* gives a more in-depth analysis of this transformation, and being a film critic, examples from the film world come in handy:

> The camera has never loved Gore, but something is going on in 'An Inconvenient Truth', and that's the other big story here. Like John Travolta in 'Pulp Fiction', Gore has come back on the scene heavier, older and a lot more likable, physically transformed in a way as to allow people to see him as if for the first time. After years of looking like Clark Kent without the glasses, Gore looks like a heavyset mensch. Moreover, the change seems to be more than surface. In 'An Inconvenient Truth', Al Gore has the look of a man who's been through something big and awful and has come out the other side.
>
> (June 2, 2006)

The comparison of Gore with Travolta's ultra cool character Vincent Vega in *Pulp Fiction* (Quentin Tarantino, 1994) has, at first view, nothing to do with ecology. Indirectly, however, it has a lot to do with just that when analysing it from the viewpoint of masculinity. In an analogous way to how the female ecoheroine has dominated the feature ecofilm, the ecodocumentary has been dominated by male film-makers whom often place themselves in the middle, Michael Moore style, of the story they convey. Although not formally the film-maker, Gore was nonetheless an indispensable part of *An Inconvenient Truth* and the discussion of his persona continually tended to slide over to that of his masculinity and how it was transformed and constructed. When Chris Knipp in *Baltimore Chronicle*, for example, describes the new Gore he enumerates a cavalcade of typical 'male' characteristics: 'calm, passionate, rational, and smart; he's not a people-pleaser that Clinton is, but may have more real warmth. Gore has become more outspoken, strong, free, his own man since his "defeat"'. (Knipp: June 2, 2006). In other words Gore had not just changed, to a high degree this transformation has also contributed to the film's ecological credibility given that he had conquered the historically important sign of character.

As David Tjeder suggests, a man's character came into prominence as a genuine expression of real manhood during the nineteenth century. The idea that a man should and could mould their own character and that the inner qualities mirrored the outward appearance has since then become the norm or the measurement of masculinity in the Western world. Gore's comparison to Clinton in the quote above is telling when Tjeder writes: 'Character was […] the opposite of feigning behavior. A man of character had his worth grounded in himself, not in the mere display of polished manners' (2003: 59). And of course, a man of character, a real man, was not only superior to other men and women, he also radiated the same inner reassuring confidence that Al Gore seemed to do according to the vast majority of reviews. Typically, the reviewers also saw this as a success story á la Hollywood: 'Between the lines, *An Inconvenient Truth* is a quintessentially American story of reinvention' (Thomson: June 2, 2006).

However, the fact that ecodocumentary of the new millennia can be gendered as predominately male is not the same as to claim that they had character. *Encounters at the End of the World* (Werner Herzog, 2007), *Food, Inc.* (Robert Kenner, 2008), *The Cove* and *GasLand* (Josh Fox, 2010) were all nominated for Best Documentary Feature after *An Inconvenient Truth*, and once again it should be noted that the ecofilm stays in the rearview mirror of other genres. During the same period the nominations were, for instance, overrun by documentaries that dealt with the Bush administration and the war in Iraq, such as *No End in Sight* (Charles Ferguson, 2007) and *Restrepo* (Tim Hetherington, Sebastian Junger, 2010), something which was also reflected in the feature categories with the big winner of the Academy Awards 2010, *The Hurt Locker* (Kathryn Bigelow, 2008).

Although the subjects of the nominated ecodocumentaries varied widely from German director Werner Herzog's existential journey to Antarctica in *Encounter at the End of the World*, to the ecological spy-adventures in *The Cove*, and the hydraulic fracture mining in *GasLand*, the main common denominator was the flagrant placement of male 'experts', crusaders and talking heads in front of the camera. In *GasLand* the inevitable comparison with Michael Moore comes up in reviews as director Josh Fox makes his documentary investigation of corporate America's abuses 'in the shadow of Michael Moore and the doorstepping stunts that broke ground in reaching bigger audiences' (Davies: February 2011) with films like *Bowling for Columbine* and *Capitalism: A Love Story* (2009). While some call this 'meticulousness' (Davies: February 2011) others wonder if Fox does not push 'his luck on the soundtrack by coming on like Martin Sheen in *Apocalypse Now*, his ultimately numbing voiceover delivered in halting rhythms and hushed tones' (Nelson: September 15, 2010).

One could pose the question if it was necessary to adopt Moore's tactics in the ecodocumentary in order to receive a precious Oscar nomination. Was the attention span of the Academy so low when it came to ecological matters that these strategies had become compulsory? If one takes the dominance of disaster and conspiracy themes in earlier feature films into calculation, then perhaps the romantic and therapeutic display of nature was not enough to stir up the necessary emotions. This becomes evident when looking at another common trait in these ecodocumentaries, namely the fact that all, with the exception of the quirky yet beautiful *Encounters at the End of the World*, showcases a nature in despair; an environment that already has gone terribly wrong under the influences of man (that is, American and Japanese corporations) and where, quite explicitly, real men are needed to make things right, or at least sound off the alarm.

In *The Cove*, the film about the unethical hunt for dolphins in Japan, former dolphin trainer now turned activist Richard O'Barry, and film-maker Louie Psihoyos, take on the role of ecoheros. O'Barry trained dolphins for the popular TV-show *Flipper* (NBC, 1964–1967) before quitting, convinced that dolphins in captivity were tortured creatures. He, and probably also executive producer of the series, Ivan Tors, was in fact responsible for the performing dolphins craze that started in the 1960s, and hence redemption is one of two themes in this documentary that connects ecology and masculinity.

Beginning with the numerous passion plays, such as American *Passion Play* (1903) and French *La Vie et la Passion de Jesus Christ* (Lucien Nonguet, Ferdinand Zecca, 1903), masculinity has, throughout film history, been connected to the concept of redemption. Secularization has, however, feminized this religious-Christian theme in a way that has disconnected it from its original connotation as something hyper-masculine (Gustafsson: 2008: 91–113). This implies that overt religious displays have, over the years, been opted out in favor of more secular versions of masculinity, for example in films like *Silent Running* and *The Omega Man*, the latter with a barely disguised Christ-figure in the form of Charlton Heston. Redemption, nevertheless, has continued to be strongly connected to the concept of masculinity. In *The Cove*, O'Barry's suffering for the environment and subsequently 'salvation' has a powerful emotional impact: 'His drooping eyes and sagging shoulders testify to the bone-deep exhaustion of someone who has spent the last 35 years atoning, and when he gate-crashes a meeting of the International Whaling Commission, the video screen strapped to his chest is like a physical manifestation of decades of guilt' (Catsoulis: July 30, 2009). Roger Ebert, who once again gave a wholly uncritical review for an ecodocumentary, was deeply moved by O'Barry's persona: 'But when all of the facts have been marshaled and the cases made, one element of the film stands out above all, and that is the remorse of Richard O'Barry' (August 15, 2009).

Ebert was not the only one that reviewed these nominated ecodocumentaries as fact-of-the-matter documents, probably as a result of the gruesome images they contained and that were described as the 'barbaric slaughter of innocents' (Travers: July 30, 2009) with 'methods [...] so nonchalantly brutal and gut-churningly primitive' that 'urbane eco-warriors' (Catsoulis: July 30, 2009) were urgently needed. The other theme was the more 'manly' genre elements of the 'activist action-movie' with the 'caper-movie touches and cocky self-awareness' (Biancolli: August 7, 2009) that director Louie Psihoyos physically represents since a lot of *The Cove* revolved around the spy-like preparations and team gathering á la *Mission: Impossible* (Brian De Palma, 1996). Although not as physical, the same approach can be found in *Food, Inc.* where the authors Eric Schlosser and Michael Pollan act as fearless intellectual ecowarriors, and director Robert Kenner, although not diegetically in the picture, tracks down and urges farmers and ecological victims of the food industry to tell the truth, while visually perusing mega multinational agricultural biotechnology corporations such as The Monsanto Company with appalling footage of animal abuse that was described as 'mind-boggling, heart-rendering, stomach-churning' (Biancolli: June 12, 2009)—even as 'medieval' (Knipp: July 7, 2009)—by a cohesive group of chocked reviewers. To conclude with another confession by Ebert:

> This review doesn't read one thing like a movie review. But most of the stuff I discuss in it, I learned from the new documentary 'Food, inc.', directed by Robert Kenner and based on the recent book *An Omnivore's Dilemma* by Michael Pollan. I figured it wasn't important for me to go into detail about the photography and the editing. I just wanted to scare the bejesus out of you, which is what 'Food, Inc.' did to me.
>
> (June 17, 2009)

Conclusion

The historical overview of ecofilms nominated for Academy Awards between 1979 and 2011 reveals both continuity and disruption when it comes to ecological themes displayed in the four cycles that have structured this article. The threat of atomic energy and nuclear contamination, which first made its impact on the members of the Academy of Motion Pictures Arts and Sciences in 1979 with *The China Syndrome*, have gradually all but disappeared for the benefit of manmade and by now ongoing threats to the environment, such as industrial pollution, global warming, overpopulation and the overexploitation of Earth's natural resources. The frequency of ecofilms have also fluctuated a great deal, with two peaks during cycle 1 (1979–1985) and 4 (2006–2011) and with a more or less coherent plunge during the period in between (1986–2005). These peaks are clearly connected to the concerns over nuclear power after the Three Mile Island accident in 1979, and to the ecological awareness of the changes in the global climate that were updated in the wake of the Kyoto Protocol in 1997. The lengthy interval is harder to explain, especially if one considers that the Chernobyl disaster happened in 1986, just at the end of cycle 1. Given that Chernobyl is considered to be the worst nuclear power plant accident in history (BBC News 2011),[14] one would think that it would have made an even bigger impact on popular culture than, essentially, a lone TV-movie, *Chernobyl: The Final Warning* (Anthony Page, 1991).

As Sean Cubitt asserts, 'we have no better place to look than the popular media for representations of popular knowledge' when it comes to mediations that concerns different environmental issues (2005: 1). But what about when a theme disappears out of sight? For sure, the nuclear energy theme did not disappear altogether, but the lack of Oscar nominations reflects the evaporation of interest in it from the film business. Simply put, the audiences' interest had been met and their worn-down attention span was not even affected by the Chernobyl disaster—which of course also took place far away in the Soviet Union. However, one should note that this, from an educational viewpoint, does not indicate that audiences suddenly forgot what the previous cycle had told them about nuclear energy and its dangers. On the contrary this knowledge works on a cumulative level as the act of repetition has the ability to affect the audience's thinking and behavior, although the actual substantiation for this didactic influence still rests upon individual examples such as the impact of *The Birth of a Nation* (D. W. Griffith, 1915) and *Holocaust* (Toplin 2002: 178–96). Likewise, one can expect that the media craze around global warming and climate change will eventually decrease, that is, if it already has not done so for the benefit of, still ecological, questions on global pandemics, as in *Contagion* (Steven Soderbergh, 2011), or the mistreatment of Earth's limited recourses as discussed in *Fuel* (Oktay Ortabasi, 2008).

Furthermore, the different cycles of ecofilms have been noticeably gendered according to specific genres. The ecoheroine emerges in semi-conspiracy melodramas, in courtroom melodramas, and as female farmers in the Country trilogy, while the ecohero can be chiefly connected to the fourth cycle's ecodocumentaries. Interestingly enough, the ecoheroine is closely connected to the melodramatic via emotions, family, and not least as the main protagonist which stands up against the main antagonist in the ecofilm, that is, the big greedy corporation which is governed by an emotionless Patriarchy. What we have, then, is a development that turns ecology and the protection and preservation of nature into something which is coded as feminine according to the notion that women, historically, have been more closely associated with nature than men. One could rightly accuse these films for essentialism when it comes to the reproduction of gender. Nonetheless, even if the female position is subordinate in a patriarchal hierarchy it is, at the same time, possible to turn this view upside-down and claim that the ecofilm in general was narrated through a feminist viewpoint, or according to Brereton's model over Light versus Dark ecology, as post-patriarchal (2005: 27).

The ecohero, on the other hand, is more closely connected to the superior concept of culture; commonly in the factual ecodocumentary with its semblance of rationality and objectivity. However, since the detached big business corporation is the main antagonist, also here the male protagonists are forced to utilize a high degree of emotion in order to resist the patriarchal corporation, and in turn, to be perceived as real men and ecowarriors (and of course in order to attract an audience). In other words, the gendering of the ecofilm, whether feature films or documentaries, meant that the ecological subject in one way or another was drenched in emotions such as passion, sentimentality, indignation and fear. This, in turn, meant that the educational value of these films was dependent on, and realized through, these different so called feminine emotions. The (biased) sentiment and affecting argumentation in itself therefore became a vital part of the ecological education, simply because of these films position as popular culture—and this in contrast to the less accessible results of environmental sciences.

> Fine art and popular media alike can, at their best, be far more than symptoms of their age. They can voice its contradictions in ways few more self-conscious activities do, because both want to appeal directly to the senses, the emotions and the tastes of the hour, because both will sacrifice linear reason for rhetoric or affect, and because both have the option of abandoning the given world in favour of the image of something other than what, otherwise, we might feel we had no choice but to inhabit.
>
> (Cubitt 2005: 2)

Despite the fact that the analysed films have been seen through the conservative lens of the Academy of Motion Pictures Arts and Sciences, the overall feminist approach in them is still strong which verifies that gender and the notion of the ecoheroine are essential for the understanding of the development of the American ecofilm.

References

Altman, R. (2004), *Film/Genre*, London: BFI.

Ang, I. (1985), *Watching Dallas. Soap Opera and the Melodramatic Imagination*, London: Methuen.

Appelby, J., Lynn, H. and Margeret, J. (1997), 'Telling the truth about history', K. Jenkins (ed.), *The Postmodern History Reader*, London: Routledge.

Bakan, J. (2005), *The Corporation: The Pathological Pursuit of Profit and Power*, London: Constable & Robinson.

BBC News, Richard Black (November 15, 2011), 'Fukushima: as bad as Chernobyl?', http://www.bbc.co.uk/news/science-environment-13048916. Accessed September 15, 2012.

Biancolli, A. (June 9, 2006), 'An Inconvenient Truth. Gore's mix of emotions and science really works', *Houston Chronicle*.

—— (June 12, 2009), 'Review: "Food, Inc.": not for the squeamish', *San Francisco Chronicle*.

—— (August 7, 2009), 'Review: the cove', *San Francisco Chronicle*.

Box Office Mojo, (October 4, 2011), http://www.boxofficemojo.com/movies/?id=chinasyndrome.htm. Accessed September 15, 2012.

—— (November 7, 2011), http://www.boxofficemojo.com/genres/chart/?id=documentary.htm. Accessed September 15, 2012.

Bradshaw, P. (September 15, 2006), 'An Inconvenient Truth', *The Guardian*.

Brereton, P. (2005), *Hollywood Utopia. Ecology in Contemporary American Cinema*, Bristol: Intellect.

Canby, V. (March 16, 1979), 'Nuclear plant is villain in "China Syndrome": a question of ethics', *The New York Times*.

—— (December 14, 1983), 'Karen Silkwood's story', *The New York Times*.

—— (December 19, 1984), 'Farmers' plight in "The Rriver"', *The New York Times*.

Catsoulis, J. (July 30, 2009), 'From Flipper's trainer to dolphin defender', *The New York Times*.

Chris, C. (2006), *Watching Wildlife*, Minneapolis and London: University of Minnesota Press.

Clover, C. J, (1992), *Men, Women and Chainsaws. Gender and the Modern Horror Film*, BFI: London.

Cohen, S. (2007), *States of Denial. Knowing about Atrocities and Suffering*, Polity Press: Cambridge.

Cubitt, S. (2005), *Eco Media*, Rodopi: Amsterdam and New York.

Davies, S. (February 2011), 'Film review: Gasland', *Sight & Sound*.

Doane, M. A. (1987), *Desire to Desire: The Woman's Film of the 1940s*, Bloomington: Indiana University Press.

Ebert, R. (March 10, 1972), 'Silent Running', *Chicago Sun Times*.

—— (April 27, 1973), 'Soylent Green', *Chicago Sun Times*.

—— (March 16, 1979), 'The China Syndrome', *Chicago Sun Times*.

—— (November 4, 1983), 'Testament', *Chicago Sun Times*.

—— (December 14, 1983), 'Silkwood', *Chicago Sun Times*.

—— (January 11, 1985), 'The River', *Chicago Sun Times*.

—— (September 23, 1988), 'Gorillas in the Mist', *Chicago Sun Times*.

—— (March 17, 2000), 'Erin Brockovich', *Chicago Sun Times*.

—— (June 2, 2006), 'An Inconvenient Truth', *Chicago Sun Times*.

—— (June 17, 2009), 'Food, Inc. First we make them miserable, and only then kill and eat them', *Chicago Sun Times*.

—— (August 5, 2009), 'The Cove. On the needless slaughter of intelligent mammals', *Chicago Sun Times*.

The Green Fuse, Adrian Harris, (October 31, 2011), 'Ecofeminism', http://www.thegreenfuse.org/ecofem.htm. Accessed September 15, 2012.

Gustafsson, T. (2008), 'I sekulariseringens skugga. Manlighet och religiös tematik i svensk och amerikansk 1920-talsfilm', *Tidskrift för genusvetenskap*, No. 3–4.

———— (2010), '*The Last Dog in Rwanda*: Swedish Educational Films and Film Teaching Guides on the History of Genocide', E. Hedling, O. Hedling and M. Jönsson (eds), *Regional Aesthetics: Locating Swedish Media*, Stockholm: National Library of Sweden.

Gentele, J. (November 17, 2005), 'Starkt verk visar ekologisk katastrof', *Svenska Dagbladet*.

Hays, M. (March 10, 2000), 'China Syndrome Lite. Julia Roberts plays evironmentalist Erin Brockovich', *Montreal Mirror*.

Hinson, H. (September 23, 1988), 'Gorillas in the Mist', *Washington Post*.

———— (November 9, 1990), 'Dances with Wolves', *Washington Post*.

Howe, D. (November 9, 1990), 'Dances with Wolves', *Washington Post*.

IAEA, (November 14, 2011), '50 years of nuclear energy', http://www.iaea.org/About/Policy/GC/GC48/Documents/gc48inf-4_ftn3.pdf. Accessed September 15, 2012.

IMDb (November 1, 2011), http://www.imdb.com/title/tt0099185/. Accessed September 15, 2012.

IMDb, nycritic, (November 6, 2011), 'Soderbergh's visuals in the service of Roberts', see also Dr Jacques COULARDEAU, 'Julia Roberts as the Archangel Gabriel' and jhclues, 'Prepare for the Oscar Roberts, Finney'. http://www.imdb.com/title/tt0195685/reviews. Accessed September 15, 2012.

Johnson, G. A. (October 14, 2005), 'Darwin's Nightmare', *San Francisco Chronicle*.

Keane, S. (2001), *Disaster Movies. The Cinema of Catastrophe*, London and New York: Wallflower Press.

Knipp, C. (June 23, 2006), 'This Man Has Something To Tell Us', *Baltimore Chronicle*.

———— (July 7, 2009), 'Robert Kenner's "Food, Inc." Big guys and little guys and what we have to eat', *Baltimore Chronicle*.

Kracauer, S. (2004), *From Caligari to Hitler. A Psychological History of the German Film*, Princeton and Oxford: Princeton University Press.

LaSalle, M. (January 8, 1999), 'Call to "Action". Travolta leads battle against polluters in fact based drama', *San Francisco Chronicle*.

———— (June 2, 2006), 'The earth is heating up like a meteor from hell and we're all going to die. Now, that's inconvenient', *San Francisco Chronicle*.

Levy, E. (2003), *All about the Oscar. The History and Politics of the Academy Awards*, New York and London: Continuum.

Macdonald, M. (June 2, 2006), '"An Inconvenient Truth": Al Gore's slide show commands attention', *The Seattle Times*.

Maslin, J. (September 23, 1988) "Saving Africa's Gorillas", *The New York Times*,.

———— (December 25, 1998) 'A Civil Action': Lawyer Errs on the Side of Angels", *The New York Times*.

McCrisken, T. and Andrew, P. (2005), *American History and Contemporary Hollywood Film*, New Brunswick and New Jersey: Rutgers University Press.

Morris, W. (March 17, 2000), 'Roberts is Sassy, Brassy in "Erin". It's a loose and funny "Norma Rae" tale of a justice-seeking legal secretary', *San Francisco Examiner*.

Murray, R. L. and Heumann, J. K. (2009), *Ecology and Popular Film. Cinema on the Edge*, New York: State University of New York Press.

Nelson, R. (September 15, 2010), 'The Case Against Fracking in *Gasland*', *Village Voice*.

Norton, B. (1991), *Towards Unity among Environmentalists*, Oxford: Oxford University Press.

PBS, (November 1, 2011), http://www.pbs.org/pov/buildingbombs/. Accessed September 15, 2012.

Puig, C. (May 23, 2006), 'Gore's "Truth" is Highly Watchable', *USA Today*.

Scott, A. O. (March 17, 2000), '"Erin Brockovich": High Ideals, Higher Heels', *The New York Times*.

——— (August 3, 2005), 'Feeding Europe, Starving at Home', *The New York Times*.

——— (May 24, 2006), 'Warning of calamities and hoping for a change in "An Inconvenient Truth"', *The New York Times*.

Soila, T. (1991), *Kvinnors ansikte. Stereotyper och kvinnlig identitet i trettiotalets svenska filmmelodram*, Stockholm: Stockholm University.

Tasker, Y. (1998), *Working Girls. Gender and Sexuality in Popular Cinema*, London and New York, Routledge.

Thomson, D. (June 2, 2006), '"Truth": More than Hot Air', *Washington Post*.

Tjeder, D. (2003), *The Power of Character. Middle-Class Masculinities, 1800–1900*, Stockholm: Stockholm University.

Toplin, R. B. (2002), *Reel History. In Defense of Hollywood*, Lawrence: University Press of Kansas.

Travers, P. (March 17, 2000), 'Erin Brockovich', *Rolling Stone Magazine*.

——— (July 30, 2009), 'The Cove', *Rolling Stones Magazine*.

Williams, L. (2009), 'Film bodies: gender, genre, and excess', L. Braudy and M. Cohen (eds), *Film Theory & Criticism*, New York and Oxford: Oxford University Press.

YouTube, (November 8, 2011), 'Republican Joe Barton v. Al Gore', http://www.youtube.com/watch?v=CqUHM2gf5g4&feature=fvst. Accessed September 15, 2012.

Notes

1 Altogether *The China Syndrome* was nominated for four Academy Awards: Best Actor in a Leading Role (Jack Lemmon), Best Actress in a Leading Role (Jane Fonda), Best Art Direction-Set Decoration, and Best Writing, Screenplay Written Directly for the Screen. The one exception when it came to Academy Award nominations for disaster films was *The Towering Inferno*, which among its six technical nomination (and three wins) also landed nominations for Best Actor in a Supporting Role (Fred Astaire) and Best Picture.

2 With the exception of short film categories in animation, live action and documentary. In addition to these, I will not consider the different honorary awards.

3 *War of the Worlds* actually won the Oscar for Best Special Effects, and was in addition nominated for both Best Sound and Best Film Editing.

4 Although the ecofilm follows cyclic behavior, we could also talk about two distinct cycles here (1979–1985 and 2006–2011) with a wide gap in between.

5 This includes nominations for Best Animated Feature for *Finding Nemo* (2003) and *Ice Age* (2004).

6 It is a matter of debate if Karen Silkwood was murdered in this car accident in 1974, or if she just fell asleep at the wheel.

7 *Building Bombs* was subsequently broadcasted by PBS on 10 August 1993, some 5 years later.

8 For criticism of melodramatic elements, see, e.g., Kracauer 2004 and McCrisken and Pepper 2005.

9 For a theoretical overview, see Doane 1987.

10 For further reading, see Ang 1985.

11 Director Steven Zaillian wrote the screenplay and had been awarded an Oscar for *Schindler's List*.

12 In fact, *The Constant Gardener* (2005) resulted in an Oscar for Best Actress in a Supporting Role for Rachel Weisz, a sort of last hurrah for the ecoheroine. Symbolically, Weisz's character (Tessa Quayle) is found murdered at the beginning of the film and her husband Justin Quayle (Ralph Fiennes) takes over the ecological relay baton.

13 For example in the French (narrated by Charles Berling) and Swedish (Gösta Ekman) versions, the life of the penguins is shown in a comical manner when it came to narration and music.

14 The Three Mile Island accident is in fifth place.

PART IV

(In)Sustainable Footprint Of Cinema

Climate Change Films: Fear and Agency Appeals

Inês Crespo and Ângela Pereira

Introduction

This chapter analyses cinematic communications about climate change. Given that films reach millions of people, we consider it important to explore film as an alternative tool for science communication, especially when it comes to controversial issues such as climate change, where high stakes, diverse values and politics intertwine with the science produced and communicated to the publics. We use the concept of 'publics' to emphasize that an audience who watches a film is heterogeneous, interpreting its messages in numerous different ways; they are 'active audiences' (see Fiske 1987: 16). Climate change is indeed a relevant issue as far as films are concerned: its impacts are not immediately grasped, felt or seen, nor are they as tangible as water pollution or waste, for example. Moreover, the assessment and causality of those impacts are at the core of substantial controversy in scientific circles and highly instrumental both at political and public levels as the adoption of measures may depend on a level of consensus about the potential challenges posed by climate change. Films bring together a set of features that may facilitate communication about climate change: they generally structure information in an attractive way, they are among the most watched media programmes in Europe and people claim to use them as legitimate sources of information about environmental issues (Eurobarometer 2002). Indeed, the form of communications influences perceptions about the issues addressed and their appropriation by the publics (Boer 2008; Foust and Murphy 2009; Morton et al. 2011; Nisbet 2009; Priest 1994; Spence and Pidgeon 2010).

The influence of the apocalyptic *The Day After Tomorrow* (Emmerich, 2004) on public perceptions of climate change has been assessed by a number of scholars in different countries. Whereas in the United States the film raised public awareness of climate change (Leiserowitz 2004), in Germany and in the United Kingdom the results indicated a drop in the perceived 'probability of climate change' (Reusswig et al. 2004;

Lowe et al. 2006). Other climate change films, such as *The Age of Stupid* (Armstrong, 2009), seem to have raised public involvement in the issue in the United Kingdom and promoted mitigation action and behavioural change (Howell 2011). As for Al Gore's *An Inconvenient Truth* (Guggenheim, 2005), it apparently raised awareness of climate change in Australia (Smith and Hargroves 2007), and triggered willingness among students in the United States and United Kingdom to reduce greenhouse gas (GHG) emissions (Nolan 2010; Beattie et al. 2011).

We argue that despite their different genres and different approaches to (re)presenting issues of climate change, many of these films structure their awareness strategies in similar ways. They begin by presenting scientific arguments, which they then develop by including frightening representations of the effects of climate change and conclude with suggestions for tackling the problem. One of the elements they use to trigger a sense of agency among their audiences are fear-inducing representations. In fact, the effectiveness of fear-inducing campaigns in triggering attitude change has been widely researched in the field of medicine (see for instance Job 1988; Leventhal 1965; Rogers & Mewborn 1976; Witte & Allen 2000). These studies suggest that fear appeals are more successful when the recommendations given to eliminate the fear threat are perceived as effective. Regarding climate change, Loewenstein and Schwartz (2010) argue that the cause of collective complacency in relation to this issue is a deficit of fear. The authors claim that climate change does not induce sufficient levels of fear because it lacks two main features which usually activate the fear system of human beings: being oriented to the present and providing tangible material for contemplation (i.e. the consequences of human behaviour are perceptible). O'Neill and Nicholson-Cole (2009) have investigated the effects of fear-inducing climate change communications on people's involvement with the issue. Their results suggest that even though fearful climate change representations focus people's attention on the issue and produce a general sense of importance, this type of communication seems to be disempowering on a personal level, leading to disengagement with the issue. Given that these types of narratives are very often presented in climate change films, it is relevant to further investigate their role in constructing a relationship between the audiences and the issues the films address.

The objective of this chapter is to further our understanding of the type of strategies, individual narratives and imaginations people construct when receiving appeals of fear and agency from these types of films. We focus on three films: Roland Emmerich's apocalyptic *The Day After Tomorrow*, Al Gore's documentary *An Inconvenient Truth* and the photo-documentary *Home* produced by the French photographer Yann Arthus-Bertrand (2009). What elements of these films evoke fear among the audiences who watch them? Do these or other elements trigger a sense of agency among those audiences? To answer these questions, we have used focus groups consisting of international graduates working in Northern Italy and discussed the thoughts and emotions of the participants during and after viewing selected scenes from the three films. While not claiming to be fully representative, the reception of these films by people from different, mainly European, countries provides a transnational dimension to this reception study. The article concludes with some thoughts on the use of films for communicating climate change.

The Three Films

The Day After Tomorrow is a disaster narrative that makes intensive use of special effects to depict the consequences of anthropogenic climate change. The film tells the story of Jack Hall (Dennis Quaid), a paleoclimatologist who forecasts a major climate shift and tries in vain to warn the vice-president of the United States. Due to global warming, the thermohaline circulation system shuts down and a series of extreme weather events begin to occur throughout the world. These changes lead to a sudden ice age with a frozen New York City, where Hall tries to rescue his son. *The Day After Tomorrow* is a disaster film, which strongly distinguishes it from the other two films examined here, and which makes its selection relevant. It is among the first big-budget films specifically addressing the issue of climate change and has grossed hundreds of millions of dollars.

An Inconvenient Truth documents Al Gore's slide show presentations around the world. The film presents the results of several scientific studies, which point to the effects of anthropogenic activities on the climate. Gore shows, for instance, pictures of melting glaciers to compare past and present conditions, as evidence of changes in the climate. The film puts an emphasis on the influence of specific interest groups (e.g. oil companies) in generating misconceptions about climate change. It intersperses Gore's explanations about anthropogenic climate change with his own life story to explain how he became interested in the issue.

In comparison to our fictional and documentary examples, *Home* is a more expansive photo-documentary directed by the French photographer Yann Arthus-Bertrand. Shot entirely from the air, it was filmed in 120 locations in over 50 countries (GoodPlanet foundation 2010). Viewing Earth from a helicopter, the film examines the problems of over-crowding of the planet and unsustainable exploitation of resources. One of the film's sections addresses the issue of climate change. The film differs from the other two in the sense that it is not a Hollywood production and its distribution was innovative: it was released simultaneously throughout the world in cinemas, television and DVD, and the full-length film was freely available in several languages on YouTube.

These three films were produced with the deliberate intention of raising awareness about environmental issues, and in particular about climate change. Roland Emmerich admitted that *The Day After Tomorrow* is a result of both political and financial motivations: on the one hand, he wanted to depict his environmental concerns; on the other, he wanted to provide an enjoyable 'popcorn' film (Revkin 2004). Al Gore claimed that *An Inconvenient Truth* was 'a tool to persuade people that we have to do something before it's too late' (Daunt 2006). Commenting on the decision to create the film *Home*, Yann Arthus-Bertrand said, '[w]hat I really want is for the people whose consumption has a direct impact on the Earth to feel the need to change their way of life after seeing the movie' (GoodPlanet foundation 2009).

Fear Narrative

According to Witte (1992), fear appeals have three central constructs: fear, threat, and efficacy. The fear emotion is elicited by a threat, which is an external stimulus either consciously or unconsciously perceived by an individual. As for efficacy, it refers to the effectiveness of a message. It relates to an individual's belief as to whether a response to a threat adequately prevents it ('perceived response efficacy'), and to an individual's perception of her/his capability to effectively carry out that response ('perceived self-efficacy' — 'can I do it?') (Witte 1992; Rogers 1975). Fear can be non-cognitive. When seeing a horror film, for instance, people may feel fear for the characters they care about (Morreall 1993).

Fear-inducing representations are present in all the films discussed here. Their depictions are diverse, as they belong to three different genres. In *The Day After Tomorrow* fear is evoked by presenting climate change as a serious global issue with unexpectedly devastating consequences. In *An Inconvenient Truth*, Al Gore invokes fear-inducing tangible experiential and emotional facts, arguments, possible futures and visual scientific outputs to enhance credibility. As for *Home*, fear is invoked by visual evidence of environmental catastrophes through beautifully photographed landscapes seen from unusual visual angles (from the sky). All of the films position their viewers to reflect on the danger of losing Earth's balance.

Fear-inducing representations comprise one of the ways in which the film-makers attempt to engage audiences with the issues portrayed and ultimately inspire ecofriendly behaviour. The films described here also evoke feelings of solidarity, cooperation and collective action. They present climate change as a global issue and remind the audiences that everyone belongs to the same community, planet, 'home'. While we acknowledge that these factors may be related to fear, for instance solidarity as fear for others, we believe these belong to a different level of appeal to human emotions. They can be considered beyond fear, a concern to which we now turn.

Agency Narrative

The three films appeal for human action to tackle climate change. They use different elements to encourage agency, be it through invocation of fear, solidarity and cooperation, as previously mentioned, or through scientific and technological discourses. It is assumed in all films that the current state of knowledge about climate change belongs to the category of 'bounded uncertainty': its outcomes are known but probabilities with regard to time and scale are less certain (Brown 2004; Refsgaard et al. 2007). In order to prevent those outcomes, the films advocate the precautionary principle, that is, the risk of extreme climatic change justifies preventive measures. Whereas in *The Day After Tomorrow* this is represented by one scientist who tries to persuade politicians to take action, in *An Inconvenient Truth*, Gore claims that there is consensus about anthropogenic climate change among the majority of scientists, though doubts remain due to lobbyist groups. As for *Home,* the discourse of certainty is supported by visual evidence of environmental catastrophes (e.g. effects on the ice caps, sea-level rise, coral reefs, etc.) and reflections on time and scale of climate change effects. Hence, in all films certainty about climate change is intended to emphasize the need for action. In fact, these films downplay scientific uncertainties and controversy.

Proposed solutions to tackle climate change are not really explicit in *The Day After Tomorrow,* while, on the contrary, they are intrinsic to the message of *An Inconvenient Truth* and *Home. An Inconvenient Truth* includes illustrations to show how individual human choices can help reduce GHG emissions (e.g. graphs; Gore's speech). The film makes an appeal for change through showing lifestyles which embody the ideas advocated and appeals to necessary values: cooperation, justice, responsible consumption and accountability. As for *Home,* it alludes to our responsibility through imagery of exemplary ecolifestyles throughout the world, which contrast with reprehensible human-shaped landscapes. These are supported by appeals to action conducted through the evocation of hope as to the effectiveness of the measures being taken.

Analytical Framework

To complement what are largely hypothetical postulations of the films' effect on audiences, we now focus on empirical research carried out through focus group discussions concerning issues arising from the framing and narratives of selected scenes from *The Day After Tomorrow, An Inconvenient Truth* and *Home.* In relation to 'fear' and 'agency', the objective was to understand what kinds of strategies, individual narratives and imaginations people construct to either embrace or reject the information they receive from these highly seductive, albeit non-interactive means of conveying information.

The focus groups were organized in line with the exploratory approach described by Calder (1977), which produces qualitative data in order to provide insights into the attitudes, perceptions, and opinions of participants. Three single-session focus groups were organized, each lasting two and a half hours, in early March 2011. Particular attention was paid to recruiting citizens from different nationalities and from different scientific backgrounds with the aim of lending a degree of cultural heterogeneity to the groups. Recruitment of participants was conducted via email invitation through the use of contacts, that is people who suggested possible participants. The email invited them to watch scenes from three films and share thoughts about these, which would provide insights into the role of the films in communicating climate change and engaging audiences with the issue. Besides appetizers and beverages, no monetary incentive was provided. A questionnaire was administered beforehand to grasp the profiles of the participants as regards to age, gender, professional background and environmental practices. The participants comprised

Table 1: Participants by nationality and gender

Nationality	Gender		Total
	male	female	
Spanish	2	3	5
German	1	3	4
Portuguese	0	4	4
Italian	2	1	3
Finnish	1		1
French	1		1
Hungarian		1	1
Turkish	1		1
Swedish/American		1	1
Brazilian	1		1
Total	9	13	22

9 males and 13 females aged from 25 to 48, all holding university degrees (3- and 5-year degrees or Ph.D.), working in Northern Italy, originally from ten different countries (further information about nationalities can be found in Table 1). Two of the focus groups had six participants and the third had ten.

There was one moderator present to conduct the sessions in English, the participants' working language. The participants' scientific backgrounds included environmental science, environmental economics, geographical information systems, educational research, food security, crisis management, architecture, computer science, and various fields of biology and physics. All the participants claimed to take an interest in environmental issues, engaging in everyday environmental practices such as waste recycling. However, eight participants did not agree that climate change was caused by anthropogenic emissions. They were evenly distributed among the three focus groups.

It should be noted for the analysis that this type of profile might have influenced reactions during the focus groups, as most participants are graduates and therefore likely to be better informed about specific scientific issues than the general public. They may also tend to be critical of information received through films, in particular blockbusters, because their scientific sources are usually of a different nature.

To initiate the discussion, specific scenes from the films were shown. These scenes illustrated issues we wanted to discuss along the narratives of fear and agency. The selected scenes encompassed the main messages of the films, how these were advertised and the elements used to justify the need to tackle climate change. The scenes contained (1) scientific research on climate change, which was used by the film-makers to enhance credibility;

(2) emotional appeals and personal episodes, intended to 'humanize' the issues portrayed; (3) evidence of climate change effects, which was likely to evoke fear; and (4) solutions proposed to tackle the issue. These scenes were gathered in 15-minute long clips for each film, appearing in the same order as in the films (a description of their content can be found in Annex A). Thus, in a sense we have proposed a different shortened storyline to include the scenes we wanted the participants to observe. In the case of *Home* even though the film addresses the broader topic of sustainability and examines irresponsible human interaction with the planet, we mainly focused on scenes describing climate change. Accordingly, we must acknowledge that showing the full-length films independently of the study context could have prompted different reactions in the participants. Nevertheless, seven of the participants had previously watched *The Day After Tomorrow*, nine *An Inconvenient Truth,* and three *Home*.

Questions designed to initiate debate after each film explored feelings and thoughts relating to the scenes shown, the message they conveyed, and the solutions proposed to tackle climate change. The involvement of the moderator was confined to providing a few guiding questions to guide the discussion among participants back on track. At the end of each session, the participants were asked to compare the three films. The guiding questions can be found in Annex B. The focus group discussions were videotaped and the tapes transcribed. Analysis of these transcripts provides the basis for our exploration of the following questions:

Which elements of the films evoked fear among participants? Given that the effects of climate change are portrayed in different ways, we investigate which elements of the scenes shown invoked concern among participants. Is fear related to the plausibility of the film's scenario? Is it related to the perceived legitimacy of the film-makers or cast? Is it linked to the participant's own life story?

Which elements of the films inspired a sense of agency among participants? The selected scenes from the films provide a range of solutions to tackling climate change. This question analyses the participants' thoughts about these in order to get insights into the type of solutions they are willing to accept and ultimately adopt. We also explore the reasons for the participants' rejection of some of the solutions presented: e.g. does controversy about the films and/or solutions propose reasons not to adopt these solutions? What individual or imagined collective justifications emerge from the conversations (e.g. responsibility, legitimacy, awareness, etc.)?

Furthermore, three weeks after the focus group sessions, we distributed an additional questionnaire (see Annex C). It asked the participants whether they had watched any of the films after our discussion, whether they had searched for further information on climate change science, costs of renewable energy technologies, or on how to become more ecofriendly; whether they had become more ecofriendly after the session; and what kind of impacts of climate change (if any) out of those depicted in the films they feared.

We expected the reactions of the participants to the scenes shown to depend on factors such as the film's genre, realism, legitimacy and credibility, the participants' previous perception and opinion about climate change, trust in the films' narrators, or relationship with environmental themes. These factors would determine their level of engagement with the messages conveyed in the scenes from the three films. The following sections present focus group discussions on the selected scenes from the films. The analysis is illustrated with quotes from the participants that support (reflect or link with) or amplify (bring new elements to) the framing of fear and agency.

Which Elements of the Films Evoked Fear among Participants?

The Day After Tomorrow

The large majority of participants considered the scenes from *The Day After Tomorrow* exaggerated. They acknowledged that the film purposefully accelerated the effects of climate change to make it more spectacular. Even though most participants did not feel engaged with the scenes shown, some reflected on the possible role the film could have for other less well-informed audiences. A number of participants expressed the opinion that the film could help raise awareness about climate change among audiences who were not previously aware of the issue. This is illustrated by the following quote:

> For me, this film could work especially with children! Because they would start talking about it and the parents or teachers would teach them … To people to open their minds and start talking about it … about climate change … (P2.9, Hungarian) (All quotations by participants have been modified for readability, with due diligence paid to original wording and intent).

On the contrary, the majority of the participants presumed that the film was produced mainly for entertainment purposes rather than to convey an environmental message. This assertion outlines the participant's judgement over the implausibility of the film's scenario:

> In American film … this same situation could be provoked by aliens or any other foe; it is not very plausible. (P2.4, Spanish)

Furthermore, this comment suggests that the reactions to the films also relate to the cultural context of the participants. In fact, a number of European participants categorized *The Day After Tomorrow* as belonging to an 'American format', and that seems to influence how they receive the film. The relationship between the nationality of the participants and the reception of these films will be further explored in the section 'Transnational audiences'.

However, for at least one of the participants the scenes from *The Day After Tomorrow* did evoke concern, specifically the scene involving Northern Americans becoming climate refugees in Mexico. These scenes show a different perspective from what she had heard before and made her realize that people in poor nations are not the only potential climate refugees:

Home only shows the impact the other refugees have on you, but actually you might become a refugee yourself. (P3.5, German)

We acknowledge that particularly in the case of *The Day After Tomorrow*, seeing only the selected scenes rather than the full-length film may have prevented the participants from being drawn into the plot. For this reason they may have felt less engaged with the issues portrayed and have also become more aware of its scientific exaggerations. While this may have affected their reactions, the commentary about the film scenes during the focus group discussions is, in fact, in line with the results obtained in the United Kingdom and Germany, where the film was considered exaggerated (Reusswig et al. 2004; Lowe et al. 2006). Similarly, the focus group participants' lack of concern or fear after watching the film scenes confirm Foust and Murphy's (2009) suggestion that framing climate change discourse in apocalyptic tragedy may diminish the perceived efficacy of humanity in tackling climate change.

An Inconvenient Truth

After seeing *An Inconvenient Truth*, the melting of glaciers and the consequent sea-level rise caused by climate change became a factor of concern among the participants, especially with one participant originally from a coastal zone in Spain. As the rise of the sea-level would affect her personally, she felt emotionally connected with this particular aspect of climate change addressed in both *An Inconvenient Truth* and *Home*. She repeatedly stated after seeing each film:

For me, these images of the water coming to the cities were scary. (P3.2, Spanish)

Another factor which impressed the participants was the speed with which climate change effects may develop. For instance, *An Inconvenient Truth* and *Home* show ice melting fast, a process that is attributed to climate change. After seeing such images in *An Inconvenient Truth*, a participant said:

For me, the image of ice melting in 35 days … I didn't really know what it was but it was something that melted in 35 days … that was quite scary. (P3.2, Spanish)

Also the possibility of an ice age induced by climate change struck some participants. However, the reactions seemed to depend on factors other than the depiction of an ice age in these films. Rather, concern was connected to the film's reliability, as this participant stated:

> In *An Inconvenient Truth*, Gore speaks about the ice that melted and interrupted the Gulf Stream. That's too close to *The Day After Tomorrow*. That's scary. They say that Europe was frozen before because of this! The first film seems very unrealistic but in the second they say that in 10 years there was an ice age ... (P2.6, Italian)

The film's genre plays a relevant role here as this participant considered Gore's film more reliable and realistic than Emmerich's accelerated scenario, and therefore he felt concerned only after seeing the more 'trustworthy' *An Inconvenient Truth*. In fact, this commentary supports the results obtained by Lowe et al. (2006) with participants claiming that the film would impact them more if it had been produced by a credible source such as the BBC. In order to be effective, a fear appeal should come from a reliable source.

Home

The participant discussion over the reliability of the models and predictions presented in cinema continued in relation to selected scenes from *Home*. The film suggests that climate change will have major effects throughout the world, and mentions different dates announced by scientists for these effects to take place. Here, it was not only the potential major effects of climate change that induced fear but indeed the uncertainty mentioned in the film about the speed at which they might occur. The following discussion took place after screening of the selected scenes:

> – I think fear arises from the uncertainty that no one really knows if things happen faster than they forecast currently. (P3.3, German)
> – Yes, but think about your kids! Even if it happens in 50 years, you're afraid, right? (P3.1, Portuguese)
> – What I mean is that they state 'maybe this level is reached in ten years or so' but no one knows if it happens in five years or two so it affects you more than if it happens in 50 years. And admittedly, it concerns your kids, of course, so maybe it is a collective social fear rather than a personal fear because it might not happen in your life time. (P3.3, German)

These two participants were considering two different scenarios. Firstly, P3.3 was concerned with the possibility of facing, earlier than had been predicted by scientists, disasters caused by climate change. Secondly, both P3.1 and P3.3 were concerned with the burden on future generations. In that case, fear arose from solidarity towards other society members who would suffer the consequences of climate change. Participant P3.1 also claimed that the solutions presented in *Home* were not the type of actions that a single individual could adopt. In this sense, she felt impotent about the possible effects of climate change. Her 'perceived self-efficacy' was low. She explains as follows:

- Did you experience feelings of fear watching the film? (Moderator)
- It raised fear but it also gave solutions to counter this fear. (P3.6, German)
- But they're not really at your level, that's the thing. (P3.1, Portuguese)
- But they gave so many examples that for sure you can find one where you can contribute. (P3.6, German)
- That's for sure, but will it be enough? That's my fear, that you can do it but ... (P3.1, Portuguese)
- ...it's not enough. (P3.6, German)

So, the 'uncertain certainty' mentioned in the film actually reinforced the participants' fears both as individuals and as society members. The selected scenes raised concerns among participants about insufficiency of available resources to tackle climate change, evoking fear of the burden on future generations and of unexpected disasters.

Which Elements of the Films Inspired a Sense of Agency among Participants?

The Day After Tomorrow

In general, the participants claimed that the 15-minute clip from *The Day After Tomorrow* watched in the focus groups did not motivate them to act in a more ecofriendly way. The participants attributed their disengagement to three main factors: (1) the film did not propose any solutions to tackle climate change, (2) the way the film depicted the effects of climate change did not give an opportunity for individual action to help solve the problem, (3) the film seemed to suggest that no matter the consequences of climate change, there would be survivors.

Starting from the first point, the participants explained that the scope of *The Day After Tomorrow* did not seem to leave room for raising awareness of climate change. Even though in the questionnaires administered before the focus group, the participants considered their behaviour to be ecofriendly, they felt the film should include advice on how to improve their behaviour, as noted in this quote:

They say we have to take care but they don't say how we should take care; I'm not sure that they are teaching anything. (P2.2, Portuguese)

As for the second point, the participants felt that their potential for contributing to tackling climate change was diminished when facing the speed with which climate change occurs.

The apocalyptic circumstances of the ending once again underline an individual's low self-efficacy. As this participant stated:

It doesn't offer solutions because climate change is already [too advanced]. All you can do is just move south. (P3.6, German)

Finally, the film's main characters survive the sudden climate shift—not surprising in a Hollywood production. After seeing the film, the participants did not feel concerned about its representation of climate change and its potential consequences because, according to them, *The Day After Tomorrow* seems to suggest that it is possible to survive even in the worst-case scenario. The following quote emphasized this situation:

And audiences watching the film will by the end think 'Okay, but there will be a solution'. (P3.4, Portuguese)

Hence, the film's exaggerations seemed mostly to disengage the participants from the message conveyed.

An Inconvenient Truth

Many participants praised *An Inconvenient Truth* for the solutions it proposes, especially the ways the film made the participants feel as part of the solution to tackling climate change. For the focus group participants, engagement with climate change issues did not depend merely on the reliability of the scientific information presented in the film. Rather, the participants seemed to consider it fundamental to have explicit information about how they could reduce their GHG emissions. The following conversation illustrates these thoughts:

- The simple sentences at the end of the film gave practical examples, for instance, 'use ecological bulbs' or 'plant trees' or 'recycle'. It's the best thing a film can do! It's clear as day. That was the most important part for me. (P2.1, Turkish)
- It's nice but for me it just goes very quickly and maybe you pick up a thing or two. But it could have been better to spend 20 minutes or so in the film explaining people how to change their lives. (P2.3, German)

It was particularly the simplicity and tangibility of the recommendations given in the final part of *An Inconvenient Truth* that engaged participants with the message conveyed. In fact, these recommendations engaged also the most sceptical participants. In these cases, the participants claimed they would embrace some of the film's proposals to ameliorate environmental conditions even if they did not agree that climate change was happening due to human emissions. Thus, the final statements of the film motivated engagement, be it for climate change or for the broader sake of sustainability.

Home

Most participants felt engaged with the ability to feel as part of the solution when watching *Home*, much as they did with *An Inconvenient Truth*. Furthermore, much of the discussion focused on one of the film's key messages: 'It's too late to be a pessimist' (Arthus-Bertrand 2009). *Home* calls for urgent action but the film conveys this message with a positive tone as it shows successful instances of measures adopted by several countries to reduce GHG emissions and mitigate climate change. In fact, most participants mentioned the film's positivity, which engaged them with its message. The following quote is actually from a climate change sceptic who nevertheless identified with the film's message:

> I like it! We are living in a wonderful planet and it may become worse, but we are doing something to keep it nice. It's a message of hope. That's not the case with Al Gore's film. (P1.1, French)

Some participants felt that *Home* called for collective action rather than focusing on the individual. The film evoked among some participants a feeling of being part of the 'Earth community', as most of them felt part of both the problem and the solution. The following participant stresses how global the problem is:

> The message is for everyone, it involves all of us. We are all part of this; we all have responsibilities in what we're doing. In that sense it is for everyone. (P3.4, Portuguese)

This quote came from a participant who claimed to be sceptical about climate change being caused by anthropogenic emissions but nevertheless the film made her feel part of a real global problem. Engagement with the film's message thus did not depend on the participant's standpoint on climate change. *Home* was thus able to trigger among several focus group participants feelings that expand to a broader human scale, a sense of the need to protect Earth as part of a collective, not only in terms of being the cause of the problem but also the most reasonable solution to them.

The imagery of the film also triggered the feeling of being a part of a wider planetary collective. A participant recalled this in the following:

> Having these amazing views of the Earth, and seeing that you may lose it, really makes you think about it. It's an analogy with astronauts that see the Earth from the space as a small fragile thing, something that we have to take care of. (P1.4, Finnish)

Some participants emphasized the importance of changing social attitudes towards sustainable forms of action. One participant in particular, who had watched *Home* three times before the focus group, stressed:

> It's about 'sharing, intelligence and moderation'. We all take airplanes; we all have our cars ... If you want to have an impact, you have to change your mind in a much more universal way of acting towards the others. If you're seeing those solar panels in the desert, the first idea is that those guys need help. We're consuming far too much: moderation! We have to use the science that provides us with a lot of solutions. (P2.8, Brazilian)

However, some participants criticized the kind of solutions *Home* proposes. According to them, these solutions were mainly concerned with providing a technological fix involving high costs, focusing on the implementation of alternative energy sources rather than on the reduction of consumption—they were not focused enough on individual agency. For instance, when asked whether they would be willing to install solar panels in their homes, the participants were concerned about the financial investment these energy sources required. When discussing the self-sustainable neighbourhood in Germany shown in *Home*, the participants wondered about the costs of renting or buying such an apartment, given the modern technologies in use. This type of solution was even interpreted by some as a form of hidden agenda meant to promote sales of such technologies. In particular, a participant from a low-income country recalled the difficulties of implementing such kinds of technology elsewhere. According to her, many countries do not possess either the climatic and/or financial conditions necessary for using specific types of renewable energy. In the following quote, a participant suspected that the technologies presented in the film (e.g. carbon capture, solar panels, wind energy) would only encourage *status quo* consumerism:

> – It is somewhat peculiar that all the solutions the film presents have to do with a technological fix. (P1.3, Portuguese)
> – I think that's positive. (P1.4, Finnish)
> – I'm not sure, because the film questions the way we choose to [get] involve[d] and then, all the solutions proposed are presented as compatible with the lifestyles we have now, only reliant on technological progress. (P1.3, Portuguese)

The solutions proposed by the film were perceived to be not for the individuals but for higher institutional levels, be it for organizations, the industry, policy-makers, or even for the higher level of the 'humankind', as noted in this conversation:

> – They show new sources of energy. (P3.2, Spanish)
> – Yes, but it's more to do with countries and institutions, it is not as if every citizen can do something about them (P3.1, Portuguese)

– Yes, they are for a higher level. I thought at the end: so this is all for politicians. (P3.3, German)

For these participants, the duality of *Home* relied on a two fold problem: on one hand, the film is successful in evoking instincts for protection of Earth, the 'home' of society; on the other, it directs all solutions to a higher level of hierarchy.

Transnational Audiences

A particularly relevant issue concerning the recruitment of participants was our aim at gathering a heterogeneous group as far as nationality is concerned. This paid off in our focus group discussion as we came to notice specific comments when the participants felt connected to the film's scenes because they reminded them of their home countries. As mentioned in the analytical framework section, the majority of participants belonged to European countries, with only one participant from the United States (with double citizenship). This seems to have influenced how they perceived the films. Many criticized *An Inconvenient Truth* and *The Day After Tomorrow* for focusing on the United States as they considered the use of examples, such as the effects of climate change on New York City, and namely on the area where the World Trade Centre was located, as a tool to emotionally engage American audiences. For these European participants, the scenes had the reverse effect as they disconnected the participants from the issues addressed, be it because they did not feel threatened by catastrophes taking place on the other side of the ocean and/or because they have perceived that the film-makers were attempting to manipulate their emotions.

Given that the participants originated from ten different countries, they were asked to reflect whether their opinion would be different if the films referred to places with which they were familiar. The German participants admitted to having felt an emotional connection when they saw the German econeighbourhood as a model to follow in *Home*. Similarly, a Portuguese participant recalled that Portugal is assessing the potential of waves to produce energy, and proudly wondered whether the images of 'wave snakes' were shot in the country. In contrast, a Hungarian participant noted that her country does not possess the necessary climatic and financial conditions to adopt the solutions proposed in the film. Two Spanish women reflected on familiar effects of climate change on Spain. Whereas the participant from a coastal area was concerned about sea-level rise, the other recalled her neighbours who had died due to a heat wave in 2003 mentioned in Gore's film. The film thus resonated in her memory as seeing familiar places and situations or even measures not applicable in their contextual realities prompted emotionally charged commentary among this international audience. This suggests that climate change communications need to be considered in a transnational framework to be more effective on a planetary scale.

Table 2: Number of participants fearing climate change effects presented in the films

Melting of glaciers	14
Others becoming climate refugees	12
More frequent and intensive droughts	12
Stronger and more frequent hurricanes	12
Flooding due to sea level rise	8
Faster spreading of diseases	8
New diseases emerging	7
Yourself becoming a climate refugee	3
Solutions to preventing damage will be created	2
I imagine them too far in the future	1

Final Questionnaire

In the questionnaire administered three weeks after the focus group sessions, six participants actually stated they had adopted new ecofriendly measures after the session, and three claimed that the reason why they did not change behaviour was that previously they were already sufficiently ecofriendly. Three participants did search for further information on how to become more ecofriendly and six learned about the costs of implementing renewable energy technologies. When asked about which impacts of climate change depicted in the films they feared, the most popular answers concerned the melting of glaciers, more frequent droughts and hurricanes, and other people becoming climate refugees (see Table 2). Three participants were afraid of becoming a climate refugee and only one participant showed concern about the possibility of a new ice age.

Given that we have not inquired about this concern in the questionnaire delivered before the focus groups, it is not possible to deduce whether it was induced by watching the films. It does allow us, however, to identify the elements of these films that are likely to induce fear in spectators.

Conclusions

The three films discussed here include fear-inducing depictions in an attempt to encourage a sense of agency among their audiences. For this, they employ different strategies. The apocalyptic approach of *The Day After Tomorrow* did not succeed in engaging the participants on climate change because it was perceived as mere entertainment. While this might be in part due to the profile of the participants who are likely to be reluctant to receive information from blockbusters, and also because they have not seen the full-length film, it is also related to the film's implausible scenario and disaster narrative.

In the case of *An Inconvenient Truth,* even though the scientific content of the scenes shown legitimated its appeals for action, this was not enough to convince the most sceptical participants. Instead, the most engaging message imparted by the film concerned its suggestions on how individuals can reduce their GHG emissions. These recommendations were appreciated by all participants no matter their standpoint on climate change. As for *Home,* the scenes shown induced feelings of hope and perceived efficacy against the potential effects of climate change. This was due to the exemplary instances of measures being taken throughout the world and also due to placing responsibility on a human scale. A considerable number of participants identified with this approach.

This chapter is a contribution to exploring the roles films perform in scientific communication and public debate. Films bring together a set of features that facilitate the communication of scientific issues. They often embed science in their content, which reaches millions of people who may shape opinions about the issues communicated. Deeper understanding of their role in this process is necessary, in particular concerning the public's reasons for either accepting or rejecting the science based messages conveyed.

These types of films very often make use of fear appeals to trigger a sense of agency among the audiences who watch them. However, these fear appeals seem to have better chances of succeeding when complemented with other elements. Those include information about effective preventive measures, credibility, plausibility, and reliability of the film, emotional connection with the issues portrayed, and an overall positive and optimistic message regarding possible solutions to counter the cause of the fear. Cinema should work as a window for further exploration of the issues it deals with, since for visual attractiveness and narrative appeal, it is obvious that films cannot be comprehensive in scientific detail and therefore cannot substitute other types of scientific sources. Yet, what is important about the use of films is that they too are reliant on values, passions, dilemmas, and assumptions about human nature that ultimately constitute antecedent and emergent societal issues that surround our existence. Their richness, we would argue, is their power to represent both the Aristotelian appeals to *pathos* and *logos,* addressing issues in a way that appeals to a range of intelligent and emotional factors. The bottom line is that *per se,* films remain aspects of a broader process of communication of science and other knowledge that needs to be carefully articulated in any effort of public debate where science and technology are reflected upon. In practice, this requires that debates are carried out with strong moderation by specialists on the issues presented, who should ensure proper contextualization of the issues with which they deal.

References

Beattie, G., Sale, L. and Mcguire, L. (2011), 'An Inconvenient Truth? Can a film really affect psychological mood and our explicit attitudes towards climate change?', *Semiotica,* pp. 105–25.

de Boer, J. (2008), 'Framing climate change and climate-proofing: from awareness to action', in A. Carvalho (ed.), *Communicating Climate Change: Discourses, Mediations and Perceptions,*

Braga: Centro de Estudos de Comunicação e Sociedade, Universidade do Minho, http://www.lasics.uminho.pt/OJS/index.php/climate_change/article/viewFile/419/389. Accessed December 22, 2011.

Brown, J. D. (2004), 'Knowledge, uncertainty and physical geography: towards the development of methodologies for questioning belief', *Transactions of the Institute of British Geographers*, 29: 3, pp. 367–81.

Calder, B. J. (1977), 'Focus groups and the nature of qualitative marketing research', *Journal of Marketing Research*, 14: 3, pp. 353–64.

Daunt, T., 2006. 'In the heat of the moment', *Los Angeles Times*. Available at http://articles.latimes.com/2006/may/14/entertainment/ca-gore14/3. Accessed December 9, 2011.

Eurobarometer (2002). *Europeans' Participation in Cultural Activities: A Eurobarometer Survey Carried Out at the Request of the European Commission, Eurostat, Executive Summary*, Brussels: European Commission. Available at http://ec.europa.eu/culture/pdf/doc967_en.pdf. Accessed October 21, 2010.

Fiske, J. (1987), *Television Culture*, London: Methuen & Co. Ltd.

Foust, C. R. and Murphy, W. (2009), 'Revealing and reframing apocalyptic tragedy in global warming discourse', *Environmental Communication: A Journal of Nature and Culture*, 3: 2, pp. 151–67.

GoodPlanet foundation (2009), *Home*: An interview with Yann Arthus-Bertrand. *Home – Un film de Yann Arthus-Bertrand*. Available at http://www.homethemovie.org/en/informations-sur-yann-arthus-bertrand/an-interview-with-yann-arthus-bertrand-co-writer-and-director. Accessed December 6, 2011.

—— (2010), FACT SHEET – *Home – Un film de Yann Arthus-Bertrand*. Available at http://www.homethemovie.org/en/informations-about-the-movie/fact-sheet. Accessed December 1, 2011.

Howell, R. A. (2011), 'Lights, camera … action? Altered attitudes and behaviour in response to the climate change film The Age of Stupid', *Global Environmental Change*, 21: 1, pp. 177–87.

Job, R. F. S. (1988), 'Effective and ineffective use of fear in health promotion campaigns', *American Journal of Public Health*, 78: 2, pp. 163–67.

Leiserowitz, A. (2004), 'Day After Tomorrow: study of climate change risk perception', *Environment: Science and Policy for Sustainable Development*, 46: 9, p. 22.

Leventhal, H. (1965), 'Fear communications in the acceptance of preventive health practices', *Bulletin of the New York Academy of Medicine*, 41: 11, pp. 1144–68.

Loewenstein, G. and Schwartz, D. (2010), 'Nothing to fear but a lack of fear: climate change and the fear deficit', *G8/G20 Magazine*, pp. 60–62.

Lowe, T., Brown, K., Dessai, S., Doria M., Haynes, K. and Vincent, K. (2006), 'Does tomorrow ever come? Disaster narrative and public perceptions of climate change', *Public Understanding of Science*, 15: 4, pp. 435–57.

Morreall, J. (1993), 'Fear without belief', *The Journal of Philosophy*, 90: 7, pp. 359–66.

Morton, T. A. Rabinovich, A. Marshall, D. and Bretschneider, P. (2011), 'The future that may (or may not) come: how framing changes responses to uncertainty in climate change communications', *Global Environmental Change*, 21: 1, pp. 103–09.

Nisbet, M. (2009), 'Communicating climate change: why frames matter for public engagement', *Environment: Science and Policy for Sustainable Development*, 51: 2, pp. 12–23.

Nolan, J. M. (2010), '"An Inconvenient Truth" increases knowledge, concern, and willingness to reduce greenhouse gases', *Environment and Behavior*, 42: 5, pp. 643–58.

O'Neill, S. and Nicholson-Cole, S. (2009), 'Fear won't do it', *Science Communication*, 30: 3, pp. 355–79.

Priest, S. H. (1994), 'Structuring public debate on biotechnology', *Science Communication*, 16: 2, pp. 166–79.

Refsgaard, J. C. et al. (2007), 'Uncertainty in the environmental modelling process—a framework and guidance', *Environmental Modelling & Software*, 22: 11, pp. 1543–56.

Reusswig, F., Schwarzkopf, J. and Pohlenz, P. (2004), *Double Impact: The Climate Blockbuster 'The Day After Tomorrow' and Its Impact on the German Public*, Postdam, Germany: Postdam Institute for Climate Impact Research (PIK).

Revkin, A. C. (2004), 'When Manhattan freezes over', *The New York Times*. Available at http://www.nytimes.com/2004/05/23/movies/film-when-manhattan-freezes-over.html?pagewanted=all&src=pm. Accessed December 6, 2011.

Rogers, R. W. (1975), 'A protection motivation theory of fear appeals and attitude change', *The Journal of Psychology*, 91: 1, pp. 93–114.

Rogers, R. W. and Mewborn, C. R. (1976), 'Fear appeals and attitude change: effects of a threat's noxiousness, probability of occurrence, and the efficacy of coping responses', *Journal of Personality and Social Psychology; Journal of Personality and Social Psychology*, 34: 1, pp. 54–61.

Smith, M. and Hargroves, C. (2007), 'The Gore factor—reviewing the impact of An Inconvenient Truth', *ECOS*, 2006, 134, pp. 16–17.

Spence, A. and Pidgeon, N. (2010), 'Framing and communicating climate change: the effects of distance and outcome frame manipulations', *Global Environmental Change*, 20: 4, pp. 656–67.

Witte, K. (1992), 'Putting the fear back into fear appeals: the extended parallel process model', *Communication Monographs*, 59: 4, pp. 329–49.

Witte, K. and Allen, M. (2000), 'A meta-analysis of fear appeals: implications for effective public health campaigns', *Health Education & Behavior*, 27: 5, pp. 591–615.

Annex A

Selected Scenes from the Films

The Day After Tomorrow

Scientific background	1. Jack Hall's research findings suggest that global warming might result in an ice-age and warns politicians about it at, e.g., the United Nations Conference. 2. Scientists (e.g. from NASA) discuss the potential causes for the severe weather events taking place worldwide. 3. The US Vice-President is unconvinced by Jack's claims and refuses to act.

Evidence of climate change	1.	Severe weather events take place worldwide: e.g. tornados destroy Los Angeles; ice-cubes fall from the sky in Tokyo; a giant wave hits New York.
Emotional/ Personal	1.	Jack's son is in New York City when extreme weather events start taking place. He advises him not to leave the city library and promises to rescue him. Jack leaves in an expedition from Washington to New York.
	2.	Jack succeeds to reach New York and rescue his son.
Measures to tackle climate change	1.	In his last speech, the US Vice-President thanks Mexico for hosting American refugees and declares that the planet's natural resources cannot be consumed without consequences.

An Inconvenient Truth

Scientific background	1.	Al Gore refers to research by different scientists, including illustrative graphs.
	2.	Al Gore contests some of the claims made by the so-called sceptics.
	3.	Assessment of peer reviewed articles showing that there is actually scientific consensus about climate change. He argues that this sort of misunderstanding is deliberately created by specific groups.
	4.	Political pressures on climate change scientists demanding a disrupted disclosure of results.
Evidence of climate change	1.	Images comparing past and present conditions.
	2.	Models of potential effects of climate change in different parts of the globe.
	3.	Specific visual examples, e.g. the drunken trees and ice melt.
Emotional/ Personal	1.	How Al Gore got interested in the climate change issue.
	2.	Gore's son.
	3.	Gore visiting the farm where he lived during his childhood.
Measures to tackle climate change	1.	A graphic shows how to reduce emissions using more efficient energy tools and sources.
	2.	Advices on everyday life behaviour which everyone can adopt.

Home

Scientific background	Statements about future predictions without mentioning sources.
Evidence of climate change	Images of several places on Earth where climate change consequences are visible, places that risk being under water in the future, and statements about, e.g., lost of thickness of the ice caps, and permafrost.
Emotional/ Personal	Images of unexplored land to argue that the Earth is much older than human existence, but that nevertheless humans have disrupted its balance.
Measures to tackle climate change	The film shows exemplary measures being taken worldwide to tackle climate change, which include zero-emission houses, solar panels, wave energy, carbon capture, wind farms, and such.

Annex B

Guiding Questions Used in the Focus Group Discussions

Questions after each film

1. What are your feelings and thoughts about this film?
2. What's the main message of this film?
3. Do you trust the scientific information the film communicates?
4. Did the film invoke fear in you?
5. Do you find in the film the inspiration for changing your lifestyles and ecobehaviour? Would you adopt the solutions for tackling climate change proposed in the film?

Final questions after the three films

1. Do you think these films can contribute to raise awareness about climate change? Why?
2. With which film did you feel more engaged with the climate change issue?

Annex C

Questionnaire Distributed Three Weeks after the Focus Group

1. Have you watched any of the films after the session?

   The Day After Tomorrow
   An Inconvenient Truth
   Home

2. After the session, have you looked for further information about:

   a. Climate change science
   b. Costs of renewable energies technologies
   c. Tips on how to behave more environmentally friendly

3. After the session, are you more environmentally friendly? Y   N
4. Do you fear any of the following possible consequences of climate change, described in the movies?

   a. Flooding due to sea level rise
   b. Faster spreading of diseases

   c. New diseases emerging

   d. Others becoming climate refugees

   e. Yourself becoming a climate refugee

   f. More frequent and intensive droughts

   g. Melting of glaciers

   h. Stronger and more frequent hurricanes

   i. New ice age

   j. Other

   k. No, because I imagine them too far in the future

   l. No, because I believe that solutions to prevent damage will be created

5. Further comments:

Envisaging Environmental Change: Foregrounding Place in Three Australian Ecomedia Initiatives

Susan Ward and Rebecca Coyle

Introduction

Global media economies mobilize finance, people, ideas and production resources from either one or many localities to create cultural commodities. These are then set loose from their production contexts, to be read, interpreted and adapted to new meanings through the prism of different but comparable cultural and environmental contexts. Milton (1996) places particular significance on global flows of people, images and ideas for the carriage of discourses that shape cultural change. She prefers the term 'transcultural' over transnational to accentuate the ability of ideas and images to cross cultural boundaries (not just political borders) suggesting that environmentalism especially, as transcultural discourse, has played a significant role in processes of globalization 'by linking individuals, communities, NGOs, companies and governments throughout the world' (Milton 1996: 171). This observation is particularly pertinent when scientists are predicting the eminent demise and collapse of the earth's biosphere. This coupling of environmentalism with the understanding of the planet as a single ecosystem not only sustains the formation of environmentalist discourse as a global phenomenon, it also supports the idea of earth's humanity as 'a single moral community' with a responsibility towards the global environment, and an obligation to contribute towards collective remedial action (Milton 1996: 177–79). In this light, climate change discourse becomes a far-reaching, mobilizing force impacting around the world, though it does not make the solution to anthropogenic climate change any less problematic because climate change can mean 'different things to different people in different contexts, places and networks' (Hulme 2009: 325).

Presented below are three media models of environmental communication that circulate to global audiences as mainstream entertainment. Each is distinctive in its creative and commercial design and delivery and in line with the historical and discursive moment of their conception in their consciousness-raising intent. McGrail (2011) suggests that a feature of the modern environmental movement is its successive 'waves' of activity, with the mid-1980s witnessing the emergence of policy discourse that focused on managing 'competing claims of the economy, environment and society' (120). Since the mid-2000s the discursive ground has shifted again from concerns 'about wilderness preservation and the depletion of resources' to anxieties over the limits of earth resources, and portents of doom based on anthropogenic climate change and computerized science modelling that predict catastrophic consequences in the not so distant future. When climate change represents such a threat to humanity, it shifts the onus from managing environmental resources to effecting appropriate social and cultural change. Accordingly all three productions dealt with here must be placed in their historical context but, even so, reflect Paula Willoquet-Maricondi's distinction of 'ecocinema' from 'environmentalist films' by being consciousness-raising and activist in intention:

> Ecocinema overtly strives to inspire personal and political action on the part of viewers, stimulating our thinking so as to bring about concrete changes in the choices we make, daily and in the long run, as individuals and societies, locally and globally.
>
> (2010: 45)

The first production, *FernGully: The Last Rainforest* (Bill Kroyer, 1992), is a story inspired by the Australian region of northern New South Wales (NSW) known as Northern Rivers and by place events—specifically local campaigns to save remnants of unlogged rainforest. The production of this animation feature, however, was largely funded and coordinated through the United States, distributed by 20th Century Fox, and dubbed in ten languages. The second is the television series *dirtgirlworld* (2009–2010) created by Mememe Productions as an animation/live action montage (accompanied by website and Facebook page) about a 'gumboot-wearing girl who grows awesome tomatoes, knows cloud names and drives a big orange tractor' (McQuillen 2011 interview), and targets the preschool child audience. Cate McQuillen and Hewey Eustuce, the creators of this content, have embraced ecological sustainability in their storylines and in their industrial practice, challenging their media partners and business associates to the same through contractual arrangements that protect the ecological integrity of their content brand. *dirtgirlworld* was developed as an Australian/Canadian co-production. It has been broadcast in over 100 countries and in 15 languages, and it too has a strong connection to the rural region of Northern Rivers.

The third production, *Project Borneo 3D: 100 Days in the Jungle* (2013), is still in production (at time of writing). It is a transmedia project (a cinematic 3D documentary feature, reality television series and web content) that centres on the progress of young 'action heroes' recruited from around the world to work with local Dayak people in saving the orang-utan by halting the deforestation of rainforest habitats in the Indonesian province of West Kalimantan, Borneo. The development of this screen content is supported financially by private investors; the Queensland Government screen agency, Screen Queensland; global

Pay-TV network National Geographic; and collaboration between Microsoft Partners in Learning and TakingITGlobal. The latter two organizations manage the associated web content designed to 'mobilize a global network of five million plus young people' (Henkel 2011 interview) in fundraising campaigns that support the reforestation programme and the development of alternative ecologically sustainable forms of agriculture as an economic substitute for palm oil. The director and key producer of this project is Cathy Henkel, who, until recently, was part of the Northern Rivers community and continues to maintain close links with the locality.

Our focus on place is deliberate. Australia as a nation cannot be regarded as progressive in generating mainstream entertainment with strong environmentalist themes. Environmental communication tends to be marginalized to oppositional documentaries, factual programming and occasional infotainment. The fact that the original creators of these storylines all have strong connections to one specific rural region in Australia is more than coincidental. Accounts of environmentalism such as Thomashow's *Bringing the Biosphere Home* place emphasis on 'a place-based perceptual ecology' on the grounds that people are more likely to be spiritually and politically motivated by their direct experience with the natural world from which they derive meaning, pleasure and identity (2002: 5–6). But this does not account for why certain people and places come to have strategic roles in environmental communication—unless 'place' is perceived as a particular concentration of networked social relations that uphold these ecological values and beliefs. It then becomes possible to grasp the relations between experience of place and the global processes involved in the mediation of ecological awareness. The production studies dealt with below illustrate how it is possible for some places to inspire representations of nature that address widespread ecological concerns via the global flows of ideas and values that link people and places around the globe.

Defining the Symbolic Significance of Place

Allen J. Scott defines localities 'as the locus of dense interrelationships' positing places and culture as 'persistently intertwined' (1997: 324). Some places, such as the cities of Los Angeles, London or Mumbai, are particularly powerful because they have evolved to become the hub of dense intra- and inter-regional networks that spread across the globe through the establishment of key organizations responsible for the commercialization and distribution of cultural commodities globally (Goldsmith et al. 2010). Included in the locational assets of these cities is the critical mass of people, businesses and cultural infrastructure; as well as artistic, bohemian and 'alternative' subcultures that move in and transform neighbourhoods to a particular localized identity. This defined place identity may be particularly attractive to cultural producers and consumers enticed by the differentiated lifestyle patterns and by a sense of belonging through the comfort and stimulation of finding like-minded people (Massey 1994; Scott 1997; Florida 2002; Gibson et al. 2002).

Contrary to the spatial bias towards cities in globalization accounts, this article converges on a rural location of Australia known as Northern Rivers. As a rural locality it is less strategically placed in terms of political or cultural power. However, as a notably subtropical and scenic location, it is the locus of inter-regional and international flows because of its place identity that makes it a destination for backpacker tourists and 'tree-change' migrants from the cities, including cultural producers and followers of counter culture movements. This history, the character and constellation of networks that intersect at this particular locus, the aesthetic and lifestyle attributes of landscape implicit to its evolving place identity, and the ways in which global/local cultural economies interact, have constructed a form of 'ecocosmopolitanism' by creating symbolic goods that interpolate the viewer as one of a world community that share common responsibilities for the wellbeing of the earth's biosphere. It is our suggestion here that this place—Northern Rivers—has been particular significant for inspiring and cultivating ecologist sentiment in the form of ecomedia products that find currency in dispersed communities around the globe; though how these symbolic goods may be then translated and assimilated depends on the particular cultural understandings and experience of the consumer.

The region we target is on the east coast of Australia. It includes the rural city of Lismore, which functions as the region's commercial and government centre, although the majority of the population (approximately 60 per cent) live in the diverse scattering of organically developed villages, small towns, 'sea-change' coastal strip developments and beachside tourist centres (including the international backpacker destination Byron Bay) located along the coast and hinterland of northern New South Wales (SGS Economics and Planning Pty Ltd 2003). Historically the region has a rural economy based on agriculture, fishing, timber and forestry which is gradually being overtaken by service industries in aged care, community services, education and tourism and by the creative industries of music, fashion, fine arts, literature and film-making. This transformation is largely due to successive waves of migration from the early 1970s onward, attracted to cheap land, the rural lifestyle, lush subtropical landscapes and beaches, and the appeal of a counter culture lifestyle (McGregor 1995). A significant and visible cross-section of this migration has been artists, musicians, writers and, more recently, film-makers and multimedia practitioners who have maintained their economic lifelines from the cities north—Gold Coast and Brisbane—and south—Sydney and Melbourne. The region has become renowned for its local festivals, craft and farmers' markets, and alternative forms of media such as community radio, independent newspapers and Internet delivery that collectively disseminate 'visions of social and cultural transformation' (Wilson 2003: 7). Dubbed the 'rainbow region', the Northern Rivers has attracted scholarly attention for the articulation of social and environmental ideals that counter mainstream practice (Easton 2010; Irvine 2003; McGregor 1995). For film-makers Cathy Henkel (*Burning Season,* 2008), and Cate McQuillen and Hewey Eustuce (*dirtgirlworld*), the reason for migrating to the region was not to 'drop out' but to create a more satisfying lifestyle perceived as economically achievable through a less materialistic way of life unencumbered by excessive mortgage repayments and by maintaining communication links with metropolitan centres to generate income (Henkel 2003: 207).

Parallels can be drawn to other places subjected to similar waves of migration which subsequently recreate and reshape place identities to a subculture's ideal image, and consequently to the role of place identity in the affirmation of morally valued ways of life. Drake draws attention to the significance of place as a 'source of creative stimuli and ideas'; where place as a unique set of physical attributes can prompt a 'subjective, imagined and emotional response' that inspires the aesthetic and symbolic dimensions of content:

> These subjective, imagined or constructed localities will be a resource of prompts, signs, and symbols as important as, if not more important than, the 'real' or objective locality.
>
> (2003: 513)

The three media models presented here are embedded in social networks that have adopted environmentalism as a life project. They are part of the region's normalization of social activism 'on behalf of the planet'. The first two projects are intrinsically inspired by landscape and local place events. The third by Cathy Henkel centres on another place—West Kalimantan in Indonesian Borneo. After 17 years residing in Northern Rivers, Cathy Henkel lives a nomadic existence dictated by her film-making career as an ecological activist, although she maintains a connection to the region personally and through networks that financially support her projects. Her personal trajectory illustrates how places are never bounded but connected by social relations to places beyond.

The media projects discussed below are defined by the discourses and technologies of their historical moment, and so are presented here in chronological order. There is a span of almost 20 years between these productions—understandably environmentalism has evolved, as have available media technologies and institutions utilized for the cause of environmental activism.

Production Model 1—*FernGully: The Last Rainforest* (1992)

Nicole Starosielski (2011) suggests that animation is a neglected area of study in environmental communication but nonetheless significant because of the format's inherent strengths in 'abstraction and simplification', which enables the expression of an environmental sensibility to function metaphorically. Starosielski nominates the production and circulation of *FernGully* as a pivotal moment in 'the cycle of innovative feature films brought about by the intersection of transnational activism, corporate environmentalism and technologies that accelerate and globalize the production process' (2011: 147). The story of *FernGully* centres on the character Zak Young, a member of a logging crew who is separated from his work unit, shrunk to the size of his indigenous guide (a fairy named Crysta), and introduced to the immersive delights of a rainforest environment. With the help of Crysta and wildlife friends, Zak saves the rainforest from logging and the ecosystem from pollution. Starosielski's historicization of ecothemed animation posits *FernGully* as the exemplar of a second wave in animation (late 1980s to early 1990s) characterized by a representative practice that depicts nature as an 'elastic, potentially interactive space' (2011: 147)—in line with contemporary ideas that reconceptualized nature as a living, dynamic force that adapts, responds and transforms in response to changes in material conditions.

The animation feature is based on a children's book by Diana Young who was inspired by the creek and rainforests on her family property near the small Northern Rivers village of Federal: 'After the boys left for school I would wander down to the creek with my writing pad and spend all day there'. The feature was produced by her then-husband Wayne Young (who was associate producer on Paul Hogan's blockbuster film *Crocodile Dundee*, 1986), who was said to 'identify strongly with the creative and entrepreneurial spirit of the Northern Rivers region' (Henkel 2006: 14). Young was active in a large number of community and environmental campaigns, including the 1979 Terania Creek blockade in nearby Nightcap Ranges. This non-violent opposition to logging ancient Gondwana rainforests generated the Australian environmental movement's first sophisticated media campaign; and the campaign victory bequeathed a 'legacy of activism' centred on the protection of endangered forests in the local region (Kelly 2003: 117).

Diana Young's book was adapted to animation by the Disney veteran scriptwriter, Jim Cox, and realized visually by ex-Disney director Bill Kroyer. It was financed and distributed through 20th Century Fox and FAI Insurance in Australia (which retained the Australian rights in return for its financial investment). *FernGully* was also one of the first feature animation films to marry computer 2D applications with the traditional painted cel techniques (Orme 1992), which enabled the creators to pay special attention to background detail (Portman 1992). To ensure pictorial realism, co-producer Peter Faiman (who was also director on *Crocodile Dundee*) led 15 animators on a 4-week excursion into the Australian rainforests on the nearby NSW/Queensland border. This resulted in a hybrid graphic form in which narrative characters are stylistically conventional to mainstream animation (for example, the forest fairy suggests European mythology rather than contemporary Australian iconography) but *placed* in a lush and luxuriant environment that is explicitly representative of Australian subtropical rainforest. Indices of place are found in the depictions of natural features such as the shady glades of strangler figs, tall timbers festooned with staghorn and bird's nest ferns, diverse fungi including luminescent varieties, vocalizing frogs with billowing throat membranes, and a particularly malevolent depiction of the leeches encountered 'during one day-long trek [...] in pouring rain' (Ryan 1992).

FernGully incorporates other locational markers: Zak's wallet identifies him as a resident of Byron Bay, and the Mount Warning of the text is drawn to the shape of Northern Rivers' prominent landmark also called Mt Warning. But the producing team's desire for naturalism is also inspired by new holistic ways of seeing and valuing native rainforests; attitudes that underlie the parable of *FernGully*. Frawley (1990) notes that the European settler stance towards rainforests had undergone a significant shift by the 1970s from 'mainly regional significance' (in its utilitarian function as a plentiful source of useful timber which, once cleared, was considered potentially fertile pastoral and agricultural land) to a new paradigm in which rainforest was appreciated for aesthetic and ecological values. Within this new frame, old growth rainforests were valued as an ancient component of Australia's landscape to be treasured and protected as part of our national and world heritage (Frawley 1990: 137). The aesthetic grandeur of tall timbered forests conformed to the ideals of eighteenth century romanticism; while the rich biodiversity in flora and fauna confirmed 'holistic understanding of ecological connectedness' that became popular from the mid-1980s (Heise 2008: 22). Thus, while one source of inspiration for the *FernGully* story was local to the Northern Rivers, resistance to the destruction of local habitats was also underpinned by a rising tide of ecological values and beliefs universally accepted at a global level.

As an animation feature *FernGully* was conceived and delivered as international entertainment. A-list actors—Robin Williams, Samantha Mathis, Christian Slater and Tim Curry—gave voice to the characters. The film was successful, as Starosielski (2011) notes, and international audiences found resonance in its advocacy for the protection of rainforests worldwide. It was the first feature film to be screened at the UN General Assembly Hall on Earth Day 1992, and was shown at the UN Earth Summit in Rio de Janeiro 2 months later. Parallels were drawn with other anti-logging campaigns in the forests of the Amazon and in the high-profile campaigns of Malaysian Borneo where local indigenous communities supported by 'the charismatic figure of Swiss environmentalist Bruno Manser' (who lived with traditional hunter/gatherer communities in the area) engaged in strategies of civil disobedience to save local rainforests and local lifestyle traditions (Brosius 1999: 38). Standing in for a string of anti-deforestation campaigns in various locations around the planet, *FernGully* served as a metaphor for contemporary environmental attitudes and as an allegory for a movement that was gaining extensive media coverage worldwide, spawning a number of documentaries and a major Hollywood feature, *Medicine Man* (John McTiernan, 1992). The image of 'pristine indigenous innocents living a timeless existence in the depths of the rainforest, as bulldozers churned toward them, devouring everything in their path' (Brosius 1999: 40) continues to live on in James Cameron's more recent film, *Avatar* (2009) in which New Age notions of Mother Earth and the concept of an interactive and interconnected planetary bio-system merges with techno-utopian visions inspired by the Internet.

Production Model 2—*dirtgirlworld* (2009)

Whereas *FernGully* was inspired by environmental politics concerned with the destruction of natural habitats, endangered species and the planet's shrinking biodiversity, *dirtgirlworld*, initiated 16 years later, was conceived in a cultural and political environment

Figure 1: The cast of *dirtgirlworld*.

dominated by concerns about climate change and the need to re-immerse humanity's cultural and industrial practice within natural ecology. The producers of *dirtgirlworld* created a programme designed to educate children (and their carers) on how to live simply, sustainably and in harmony with nature; and took this political objective to a practical level by 'greening' their industrial practice and influencing their business associates to do the same based on the corporate rationale of protecting the integrity of the content brand.

dirtgirlworld, a television series made for preschoolers by Mememe Productions, first went to air on the Canadian public broadcaster CBC, (October 2009), closely followed by BBC1 and the BBC's preschooler's channel CBeebies in Britain, and the ABC in Australia (December 2009). The global headquarters of this programming content is a small converted weatherboard church near a rural whistlestop called Whiporie, located south of Casino, northern New South Wales, and the *dirtgirlworld* message of living sustainably, simply and close to nature is inspired by the lived reality of its producers Cate McQuillen and Hewey Eustace. The 52, eleven minute episodes are a fusion of live action, computer animation and photomontage where the main character, Dirtgirl, wears rainboots and grows her own vegetables. Alongside Dirtgirl are her best buddy Scrapboy, Ken the super stunt weevil, his number one fan Grubby and a monosyllabic scarecrow called Hayman. These characters perform simple storylines designed to demonstrate an ethical and ecologically centred existence. Like *FernGully*, visual aesthetics are inspired by place:

> Our blue sky we take for granted is in *dirtgirlworld*. I look up out of the window now, right at this minute, [and] look at the sky colour, then look at the poster of *dirtgirlworld*;

and it is exactly the same colour. It wasn't until I started travelling to sell the show that I realized that not all the world has a blue sky, or a purple sky, or all the other colours that we get on the North Coast. … The lightning that you can see from outside our house; the pink and blue flushes—the things we have experienced since moving here 20 years ago—to walk outside at night and see a moonbow over your house after it rains… it is just nature in the making. But if you don't see it, how do you explain it?

(McQuillen 2011, interview)

In *dirtgirlworld*, child viewers are immersed in colourful and bountiful environments riotous with flourishing flowers, plants and vegetables, and introduced to a benign micro-world of insects. Viewers learn how to farm worms or engage in Dirtgirl's problems in living ethically with nature, such as how to harmoniously coexist with the slugs that devour her lettuces. To emphasize the authenticity of their environmental message, commissioning editors and major production partners have stayed as guests on the Whiporie property 'and seen that this is a genuine real place'. (McQuillen 2011 interview)

Australia has a history of children's storytelling along environmental themes, so children's content that takes a moral and ethical stance towards the environment is not radically new. However, for Mememe Productions, the producers 'walk the talk' to the point where company ideals and brand identity necessitate close attention to the entire life cycle of their product. A dedicated green officer interrogates each line item in the production budget to find the most effective environmental solution based on three organizational objectives: 'to reduce,

Figure 2: The world of *dirtgirlworld*.

reuse, and recycle; to avoid plastic, polystyrene and PVC; and to watch consumption on all levels' (Mememe Productions 2010). A three per cent carbon offset was spent on microfinance in developing countries. For the most part, 'going green' bought substantial cost savings that enabled the company to cross-subsidize the more expensive environmental alternatives. In 2010, the production company won the Australian Directors' Guild inaugral GRASS (green awareness encouragement) award for its achievements in greening industrial practice, and it maintains the integrity of this corporate identity by exercising governance over merchandise marketed under the *dirtgirlworld* brand. In all *dirtgirlworld* contracts—whether in co-production arrangements, distribution or merchandising—'we have environmental clauses saying "for the good of the planet"' (Webb 2011 interview); and to ensure these ecoobjectives are met Mememe Productions employs a 'brand guardian':

> As a brand manager my job would be to turn the show into as much money as I can, but as a brand '*guardian*' I am still aiming to try to make money out of this brand, but I am motivated by not only making money, but doing it in the right way.
>
> (Webb 2011, interview)

Typical brand management functions, such as developing a style guide and dictating appropriate use of the trademark logo, are accompanied by ecobrand criteria, three of which are mandatory for any potential licensee or distributor wishing to secure a business partnership with Mememe Productions. The first criterion is that all products must be sustainable, with 'organic and renewable materials' given priority. The second criterion is that all packaging must be minimal, preferably re-useable and either made from 100 per cent recycled materials or 100 per cent recyclable. Madman Entertainment was chosen as the Australian distributor of *dirtgirlworld* DVDs on the basis of their promised commitment to adhere to these principles, despite the system of (red tag) security in larger retail stores that demands robust packaging. While initially sceptical of finding viable alternatives:

> [...] with a couple of gentle nudges the production team did find 100 per cent recycled polypropylene cases for the DVDs. Not only did they find them – and it hardly cost anything more than virgin plastic ones – they found compostable shrink wrap to go around the outside of the DVD case. And they were already committed to printing the sleeve for the DVD on recycled paper.
>
> (Webb 2011, interview)

The Australian distribution company was also compelled to publish their trade catalogue publicizing the launch of the *dirtgirlworld* DVD on recycled paper with accredited green printing, and have continued the practice ever since. In the United Kingdom, BBC distribution located a biodegradable DVD case and organic seeds for its DVD promotion. The third mandatory principle is that all products have to be people friendly—ideally Certified Fairtrade or at least complying to International Labour Organization standards by which companies must ensure 'there's no child labour, there are no extended hours, people aren't sleeping on factory floors, that it's not a sweat shop' (Webb 2011 interview). All potential licensees are interrogated on the source of their merchandise:

> [...] some prospective licensees say they have absolutely no idea and we have lost a couple of licensees like that; they are 'great products' but the people who are sourcing them, and calling them 'eco', really have no idea who's making them or what they are made from. And that's a bit scary.
>
> (Webb 2011, interview)

Mememe Production's media model of eco branding is unique in the Australian context where screen industry initiatives to establish sustainable practices are still nascent. By adopting the practice of 'life cycle analysis', the company has aligned its corporate identity with environmental movements in green consumerism that attempt to re-immerse industrial and market practices in the natural ecology, 'tracking' where their supplies come from, how they have been produced and by whom (De Neve et al. 2008: 7). It is a process that considers the whole commodity life cycle from product design to its end-of-life disposability and potentially it is a process that gathers momentum as more appropriate technologies, innovations, social awareness and questioning of habitual practice come online. As Barnett et al. (2005), Littler (2009), De Neve et al. (2008) and Lewis (2008) have noted, practices in ethical and green consumption have become the new terrain for political action. Like *dirtgirlworld*'s message of living simply and sustainably, it enables the utopian vision of a less exploitative relationship with the earth's ecology and resources, and relies on alternative moral perceptions of how a market economy can or should function (De Neve et al. 2008: 17). This content functions as commodity branding in the global media market but, according to producer Cate McQuillen (2011 interview): 'We are not [just] creating a brand, we are creating a lifestyle; and supporting a lifestyle that builds on people's dreams and hopes'.

Dirtgirlworld, like its predecessor *FernGully: The Last Rainforest,* is a cultural commodity 'set loose' across diverse channels of delivery according to whatever regional distribution agency decides to pay the licence fee that allows them to disseminate the material. These distribution agencies do so on the perceived 'fit' between the licensed content—particularly its aesthetic and symbolic values—and the perceived values of the targeted consumer. As entertainment that primarily targets the child consumer, both programmes have incorporated ethical and environmental principles as a way of contributing to the social and intellectual development of the child viewer, though it also ensures the shelf-life of the programme brand. However in *dirtgirlworld*'s example the production company has created an accompanying website and Facebook page that enables viewers (especially carers) to extend their engagement with the programme's message of sustainability, by accessing accredited ecomerchandise sold under the *dirtgirlworld* brand, and through further suggestions on sustainable lifestyle practice. The use of social media also provides opportunities for interactive conversations based on a shared interest in sustainability, such as posting show-and-tell photographs that celebrate audience members' initiatives inspired by *dirtgirlworld* examples. Our last production model however—*Project Borneo 3D*—introduces a new media exemplar for environmental activism through an architecture of cross-platforming content across cinema, television and the web.

Production Model 3—*Project Borneo 3D: 100 Days in the Jungle*

Cathy Henkel's multiplatform project appears to be a sequel to her 2008 documentary, *The Burning Season,* which followed the attempts of a young Australian carbon-trading entrepreneur to save Indonesian rainforests from palm oil cultivation. Achmadi, an ancillary character in the earlier film, is a small-scale farmer who lives in a village that is utterly dependent on the local palm oil industry, and the film documents his need to secure the economic survival of his family or find some other (as yet unidentified) means that is less destructive to wildlife habitats and the environment. Distributed through National Geographic International, *The Burning Season* attracted international circulation, and public and critical acclaim at festivals in Australia and the United States.

Project Borneo 3D is an ambitious project that (at May 2012) entails a feature documentary in post-production and the sequel TV series currently being shot in Borneo. The feature has an estimated budget of AUD$6 million, and a Hollywood cinematographer Don McAlpine (*Moulin Rouge, Wolverine, Medicine Man*) at the helm as director of photography. Its cinematic release is scheduled for March 2013 in North America and Australia, followed by Asia and Europe. National Geographic Cinema Ventures has secured release for the North American market and an IMAX global release and Virgo has secured distribution partners in Australia and Asia. In a pattern typical of entertainment franchises, the cinema release is intended to act as the precursor to a reality TV series by setting up the premise for viewers to follow via the National Geographic network, and/or online through a mosaic of web content created in partnership with Microsoft Partners in Learning and TakingITglobal. These latter two organizations are educational initiatives that provide teaching materials and webinars for middle to high school education. Through these organizations' associations, *Project Borneo 3D* is effectively linked to a pre-existing network of 80,000 school children in Asia Pacific classrooms. When the documentary feature has its cinema release the plan is to make this online content (designed in the form of an educational program titled DeforestACTION) available to a wider web community of five million school children that fits with curriculum in places such as Australia 'where there is an imperative to teach how students can take action in the real world' (Henkel 2011 interview).

Henkel envisages her documentary as a 3D 'action movie' to correlate with popular Hollywood ecoaction films such as *Avatar* and new entertainment standards in 3D cinema technology. But 'action movie' also refers to the narrative focus on environmental action that calls on audiences to *act* through its web of IT networks in support of the project's ecowarrior teams who are striving to save surviving remnants of rainforests in the Indonesian province of West Kalimantan in collaboration with local Dayak communities and scientists (www.

anactionmovie.com). Subscribers to the educational program DeforestACTION are invited to participate in webinars presented by members of the Borneo team, and are encouraged to be 'Earthwatchers'—'a real life youth army'—who monitor illegal logging activities online through satellite surveillance set up by the ecologist Willie Smit and Dutch company Geodan (Virgo Productions 2011). This initiative recently won the overall Top Ideas prize at the European Global Monitoring Competition (Global Monitoring for Environment and Security 2011).

Henkel describes her vision as a 'feel good story' that closely follows the allegories of *FernGully* and *Avatar*:

This is the last stand of the Dayaks. That's why I say this is the *Avatar* story that's real… The Dayak community, who are like the Na'vi, have the knowledge of the forest, live in harmony with the forest, and rely on it for their food and livelihood. They want to protect it not only for their sake but for the sake of the planet. Then you've got the scientist—like the Sigourney Weaver character—in Willie Smits; who is seen as their "saviour" because he has knowledge of alternatives to the palm oil industry. One of his solutions is development of sugar palm which supports an agroforestry system of cultivation—you don't have to chop down the forest to grow it. It is also a biofuel [and] a much better product for growing in Borneo. He also has a visionary plan for the re-growing of forest ecosystems and the rescue and rehabilitation of orangutans. His vision is very expensive and he needs these young eco warriors to help him implement the plan.

(Henkel 2011, interview)

Henkel enlisted the 'ecowarriors' using viral marketing of a recruitment clip uploaded on YouTube that called for young 'action agents' (between the ages of 18 and 35 years old) 'to save the planet' (YouTube, 2013). Over rallying orchestral music, viewers are interpolated as young people with 'the power to stand up and work for something we believe in', with the narrator calling for those who have the courage to enlist and be 'the voice, eyes and ears' of the DeforestACTION campaign.

DeforestACTION received 215 applications from 26 countries as dispersed as Africa, America, Asia and Australia in the form of 90-second camcorder pitches uploaded to TakingITGlobal's DeforestACTION website. In September 2011, the chosen 15 spent the first 20 of their 100-day enlistment journeying to remote villages in the heart of Borneo to be inducted into the lifestyle, customs and language of the local Dayak people and to see first-hand the extent of environmental and social degradation. Simple photographic diaries of the trip have been posted by Henkel on the official Action Movie website and a webinar has been conducted between representatives of the ecowarriors, the leading ecologist Willie Smits, and the 80,000 school students in the Asia Pacific region. Twitter feeds also keep followers of the website up to date on story developments. As Henkel explains:

[…] the first twenty days is like act one of the movie. At the end of the twenty days they will know what is at stake […] and the audience of the movie will know who they are, why they are there and what has to be done [...] Then they will come back; but they will have had four and a half months to prepare with resources and networks for when they come back in February to implement their plan [...] and we will bring our crews to film those 80 days of them doing it.

(2011, interview)

This first intake of recruits will become the subject of the documentary feature made for cinema release but, by the time the documentary is released to cinemas, a second intake of recruits bringing support and further funds will be introduced at the end of the feature to instigate the reality TV series.

In many ways Henkel's media project reflects 1970s' politically based initiatives such as Social Action Broadcasting in the United Kingdom or Challenge for Change in Canada which evolved from the perceived potential of television to educate and mobilize viewers around particular issues (Dovey 2000: 86). In this case, however, social networks and IT technology supplement the television product. Political action is popularized by references to 'action movies', 'reality TV', and the cross-platforming of content onto the web, formats that have currency with today's youth audience. It also signals the ability of this kind of content to generate high levels of 'connectedness' between the consumers of content and content producers who, in the commercial domain, more typically take advantage of this attraction to generate additional revenue streams through merchandising or voting on participants' status (Patino et al. 2011). Consumers of Henkel's content, on the other hand, will be asked to raise funds and donate to the restoration of destroyed ecosystems and to the development of alternative forms of economic sustainability for disadvantaged communities in West Kalimantan. Observational footage will be interspersed with 'confessional' video diaries by the ecowarriors; and to ensure this sense of intimacy between audiences and ecowarriors, Henkel will foreground three of the 15 candidate recruits to function as lead character and two supporting characters—'whose journey we follow'; focusing on those recruits who demonstrate 'the biggest potential for personal growth. It is not going to be the most confident, or the most self-assured ecowarrior. It is going to be one who has the most to learn and who will develop the most as a character on screen' (Henkel 2011 interview).

Jon Dovey (2000) observes how much factual entertainment based on the subjective experience of 'ordinary' people has become a significant part of the mix in light entertainment seen on primetime television. Under the umbrella term of 'first person media' he defines these formats as a combination of *cinéma vérité*-style or 'observational actuality footage' and 'eye-witness testimonies'. These are then edited to an overarching narrative that conforms to conventional fictions—in this case the *FernGully* allegory (repackaged and re-invigorated as *Avatar*)—with 'an overarching narrative address in which we are given to understand that our viewing pleasures are sanctioned by an explicit appeal to some communitarian logic'. Accordingly *Project Borneo 3D* appeals to the notion of individual moral responsibility to actively engage in the restoration of the earth's biosphere and to address social justice issues that recognize the destructive impact of global market forces on less advantaged communities (80–81). Dovey attributes this foregrounding of 'individual subjective experience' (as 'guarantor of knowledge') to the market-based economics of postmodern cultural economies in which texts are created around notions of pleasure and desire:

> First person media, in its constant iteration of 'raw' intimate human experience, can be seen as creating a 'balance' for that lack of narrative coherence [due to the apparent complexity and powerlessness that typically characterizes the modern condition] and for the complexity in our own lives. Subjectivity, the personal, the intimate, becomes the only remaining response to a chaotic, senseless, out of control world in which the kind of objectivity demanded by grand narratives is no longer possible.
>
> (2000: 26)

Moreover, it is this prioritizing of 'subjectivity' over narrated 'more distanced supposedly objective "truth" statements' that enable media texts as cultural commodities to more easily cross cultural boundaries. Through 'the voice, eyes and ears' of its ecowarriors, *Project Borneo 3D* offers on-the-ground understandings of the scientific, social and economic processes that lead to climate change at a particular place through the recruits (who offer points of identification for audiences) and their interaction with scientists and local Dayak people. Ultimately Henkel's vision is to offer audiences a solution and a message of hope 'as a verb with its sleeves rolled up for action' that is equally founded on notions of a unified, global moral community underpinned by the programme's web of IT networks (Virgo Productions 2011: 2).

Mainstreaming Environmentalism

In charting the shifting beliefs and attitudes towards nature mediated by media economies, Cynthia Chris (2006) notes the rise of environmentalism alongside the growth of global networks such as National Geographic, Discovery and their periodic spin-offs in niche channels of infotainment such as Animal Planet (which launched Australian television conservationist Steve Irwin into celebrity status). Along with other critics, Brockington (2008a, 2008b) and Vivanco (2004), she questions the long-term effectiveness of 'non-local environmentalism' expressed in cinema and television content. These critics note how environmental programming cash-in on audiences' emotional response to the aestheticization of nature and to reported threats of habitat loss, yet they offer little in the way of information on the political and economic processes responsible for biodiversity decline, nor explore possible social activist solutions. This is the outcome they suggest of media content that is produced as a commodity 'made to fit a market that thrives on conflict that melts into happy endings, and drama that does not get mired in real-world political impasses but resolves in comfort' (Chris 2006: 201). Her argument suggests that, in the competition for audiences and advertisers, commercial imperatives continue to shape the commodification of environmentalism according to specifications and needs of capitalist enterprise.

Dovey (2000) and Turner (2006), however, are more circumspect in their cynicism towards media's social relations of production, suggesting that, in the saturated conditions of today's media landscape, there are significant ideological consequences in this drive to extend a network's 'social and political power' in pursuit of profit (Turner 2006: 160). This is one of the inherent contradictions of modern capitalism that, in the drive to reach audiences, corporate interests may well collude with independent producers who challenge the status quo. Climate change is a phenomenon that affects everyone. According to Hulme (2009: 325) it impacts on 'our collective and personal identities' and consequently the kind of projects that do 'form and take shape'. It provokes 'new courses of action and new ways of seeing things' impacting on 'businesses, governments and law, knowledge and innovation, development and welfare, religion and ethics, and the creative world of art, cinema and literature' and on the desires and dreams of audience globally who are all troubled by the uncertainty of our future (Hulme 2009: 325).

We see evidence of this in the organizing principles of content delivery around environmentalism, and in the construction of audiences around content that inspires personal and political action. Whereas *FernGully* was exhibited in the exclusive environmentalist forums of the UN, today *dirtgirlworld* featured as one of a number of programs screened in celebration of Earth Day. A *dirtgirlworld* marathon (17 episodes aired back-to-back) featured on the dedicated pre-schooler's channel Sprout, mirroring similar programming strategies on other channels—Nickelodeon, PBS Kids, CNBC, PBS and Discovery—and in the United Kingdom, Canada and France where *dirtgirlworld* also featured in the commemoration of Earth Day (Easton 2010). A recent addition to Discovery's stable of niche channels is Planet Green, launched in 2008, that combines nature programming with lifestyle infotainment devoted to ethical consumerism. The channel pitches itself as 'the entertainment brand that champions the visionaries who move our world forward in small and large ways', and engages in ongoing 'conversations about sustainability' through two websites—planetgreen.com and TreeHugger.com. In August 2011, the channel featured a month long line-up of programmes in partnership with environmental advocacy and conservation organizations—the Alliance for Global Conservation and the Ian Somerhalder Foundation (Press Release 2011). Measuring the ideological consequences of these media productions may be an intriguing prospect (though outside the scope of this study); but definitively proving the impact of ecomedia on consumer behaviour will always be problematic. On the other hand, it is evident that with two of the media productions discussed here, there is an engagement by broad and dispersed audiences, suggesting that they have some resonance in people's lives in their awareness of a common predicament. National Geographic Entertainment's support for *Project Borneo 3D* also points to a corporate collusion in the delivery of content that conforms to shifting normative values and evolving belief systems shaped by climate change discourse. Ironically these shifts are more evident in the organization of US and European media markets, and less so in the Australian media environment where they are sporadically evident in some children's television programming, and occasional drama content commissioned by cash strapped public broadcasters, and in the odd lifestyle programme that offers environmental and ethical alternatives in material consumerism (Bonner 2011; Lewis 2008).

Conclusion

In the above we have highlighted the significance of place because of its normative function in reinforcing ethical values and environmental activism. Northern Rivers is a place that has inspired forms of ecomedia that are not common to mainstream entertainment designed expressly for an Australian market. These ecomedia projects are mediated constructions of lived environments, linked in a variety of ways to a specific Australian location, yet they are designed expressly as global entertainment and are supported by evolving trends in international markets. As examples of ecomedia, these projects draw from particular historical contexts in their 'perceptions of nature and of environmental issues' (Willoquet-Maricondi 2010: 45). *FernGully* offers 'a lyrical and contemplative style that fosters an appreciation for ecosystems' indicative of late 1980s' environmental movements underpinned by concerns for the destruction of native habitats and the inevitable demise of biodiversity (2010: 45). The more contemporary projects—*dirtgirlworld* and *Project Borneo*—deploy 'an overt activist approach to inspire our care, inform, educate, and motivate us to act on the knowledge they provide' (2010: 45) in direct response to climate change discourse and what that implies in future scenarios. While ecocinema suggests a single media format, the more recent productions examined above incorporate transmedia communication techniques to mobilize environmentalist content across a variety of platforms, including social media as appropriate to contemporary youth culture.

Ostensibly these productions target children and young people, enabling them to be fantastic in narrative, didactic in orientation, and future-focused in outcomes. Nevertheless, in the reception and circulation of *FernGully* in pro-environmental forums, in the take-up of sustainability activities designed for child play advocated by *dirtgirlworld* for child carers, and in Henkel's (and National Geographic's) expectations of a broad cinematic audience for her *Project Borneo 3D*, there appears to be a wider appeal in these messages of hope in a pro-activist stance. These three media models all posit practical 'feel-good' solutions to environmental concerns, with *dirtgirlworld* and *Project Borneo 3D* in particular offering visions of possibility to audiences increasingly anxious about projected climate change futures but based on a reframing of the environment and the individuals' place within it. Just as the early anti-logging campaigns of the Northern Rivers' Terania Creek blockade learned to use sophisticated marketing techniques of the advertising world, these projects harness corporate practice, methodologies and technologies honed by commercial interests. In doing so, they offer alternative moral perceptions on how media economies can engage with the issues of climate change, offering practical solutions and the sense of a global moral community collectively working towards an alternative future.

References

Barnett, C., Cloke, P., Clarke, N. and Malpass, A. (2005), 'Consuming ethics: articulating the subjects and spaces of ethical consumption', *Antipode*, 37: 1, pp. 23–45.

Bonner, F. (2011), 'Lifestyle television: gardening and the good life', in T. Lewis and E. Potter (eds), *Ethical Consumption: A Critical Introduction*, London and New York: Routledge, pp. 231–43.

Brockington, D. (2008a), 'Celebrity conservation: interpreting the Irwins', *Media International Australia*, 127, pp. 96–108.

—— (2008b), 'Powerful environmentalisms: conservation, celebrity and capitalism', *Media, Culture and Society*, 30: 4, pp. 551–68.

Brosius, J. P. (1999), 'Green dots, pink hearts: displacing politics from the Malaysian rain forest', *American Anthropologist*, 101: 1, pp. 36–57.

Chris, C. (2006), *Watching Wildlife*, Minneapolis: University of Minnesota Press.

De Neve, G., Leuetchford, P. and Pratt, J. (2008), 'Introduction: revealing the hidden hands of global market exchange', *Research in Economic Anthropology: Hidden Hands in the Market: Ethographies of Fair Trade, Ethical Consumption, and Corporate Social Responsibility*, 28, pp. 1–30.

Dovey, J. (2000), '*Freakshow: First Person Media and Factual Television*', London, UK & Stirling, VA: Pluto Press.

Drake, G. (2003), '"This place gives me space": place and creativity in the creative industries', *Geoforum*, 34, pp. 511–24.

Easton, A. (2010), 'Local show will go global on Earth Day', *The Northern Star*, Lismore, p. 4.

Florida, R. (2002), *The Rise of the Creative Class: And How It's Transforming Work, Leisure, Community and Everyday Life*, New York: Basic Books.

Forest, B. (1995), 'West Hollywood as symbol: the significance of place in the construction of a gay identity', *Environment and Planning D: Society and Space*, 13, pp. 133–57.

Frawley, K. J. (1990), 'An ancient assemblage: the Australian rainforests in European concepts of nature', *Continuum: The Australian Journal of Media and Culture*, 3: 1, pp. 137–67.

Gibson, C., Murphy, P. and Freestone, R. (2002), 'Employment and socio-spatial relations in Australia's cultural economy', *Australian Geographies*, 33:2, pp. 173–89.

Global Monitoring for Environment and Security (GMES) Press Release (2011), ESA's Earth Observation Director to unveil first GMES Master, 18th October. http://www.gmes-masters.com/news/esa%E2%80%99s-earth. Accessed November 1, 2011.

Goldsmith, B., Ward, S. and O'Regan, T. (2010), *Local Hollywood: Global Film Production and the Gold Coast*, St Lucia, Queensland: University of Queensland Press.

Heise, U. K. (2008), *Sense of Place and Sense of Planet: The Environmental Imagination of the Global*, Oxford: Oxford University Press.

Henkel, C. (2003), 'Emerging screen industries in the northern rivers region: a documentary film-maker's perspective', in H. Wilson (ed.), *Belonging in the Rainbow Region: Cultural Perspectives on the NSW North Coast*, Lismore: Southern Cross University Press, pp. 207–26.

—— (2006), 'Imaging the Future 2: screen and creative industries in the northern rivers region—development trends and prospects for the next decade', QUT eprints, http://eprints.qut.edu.au/4890/. Accessed October 18, 2011.

—— (2011), Interview conducted by Susan Ward, at Mullumbimby, NSW, Virgo Productions, 26 August.

Hulme, M. (2009), *Why We Disagree about Climate Change*, Cambridge: Cambridge University Press.

Irvine, G. (2003), 'Creating communities at the end of the rainbow', in H. Wilson (ed.), *Belonging in the Rainbow Region: Cultural Perspectives on the NSW North Coast*, Lismore: Southern Cross University Press, pp. 63–82.

Kelly, R. (2003), 'The mediated forest: who speaks for the trees?', in H. Wilson (ed.), *Belonging in the Rainbow Region: Cultural Perspectives on the NSW North Coast*, Lismore: Southern Cross University Press, pp. 101–20.

Lewis, T. (2008), 'Transforming citizens? Green politics and ethical consumption on lifestyle television', *Continuum: Journal of Media and Cultural Studies*, 22: 2, pp. 227–40.

Littler, J. (2009), *Radical Consumption: Shopping for Change in Contemporary Culture*, Berkshire: Open University Press.

Massey, D. B. (1994), *Space, Place and Gender*, Minneapolis, USA: University of Minnesota Press.

McGrail, S. (2011), 'Environmentalism in transition? Emerging perspectives, issue and futures in contemporary environmentalism', *Journal of Future Studies*, March, 15: 3, pp. 117–44.

McGregor, C. (1995), 'The beach, the coast, the feral transcendence and pumpin' at Byron Bay', in D. Heardon, D. Horne and J. W. Hooton (eds), *The Abundant Culture: meaning and significance in everyday Australia*, St Leonards, NSW: Allen & Unwin, pp. 51–60.

McQuillen, C. (2011), 'Interview conducted by Susan Ward', at Whiporie, NSW, 13 April.

Milton, K. (1996), *Environmentalism and Cultural Theory Exploring the Role of Anthropology in Environmental Discourse*, New York: Routledge.

Murray, R. L. and Heumann, J. (2011) *That's All Folks? Ecocritical Readings of American Animated Features*, Lincoln (NE): University of Nebraska Press.

Orme T. (1992), '"FernGully": Can a cartoon save the world?' *The Salt Lake Tribune*, 10 April. Retrieved from FACTIVA May 11, 2011.

Patino, A. Kaltcheva, V. D. and Smith, M. F. (2011), 'The appeal of reality television for teen and pre-teen audiences: the power of "Connectedness" and Psycho-Demographics', *Journal of Advertising Research*, 51: 1, pp. 288–97.

Portman, J. (1992), 'Ferngully: rainforest film "blatantly environmental", which is why big stars voiced their support', *Kitchener-Waterloo Record*, 15 April, p. E1. Retrieved from FACTIVA May 11, 2011, Portman 1992.

Press Release (2011), 'Planet green dives headfirst into its third Blue August Campaign', *Targeted News Service*, Silver Springs: Discovery Communications.

Ryan, J. (1992), 'Save the rain forest, he says, animatedly', *The San Diego Union-Tribune*, 11 April, p E-6. Retrieved from FACTIVA May 11, 2011.

Scott, A. J. (1997), 'The cultural economy of cities', *International Journal of Urban and Regional Research*, 21: 2, pp. 323–39.

SGS Economics and Planning Pty Ltd (2003), *Overview of Economic Strategies: Northern Rivers Region*, Northern Rivers: Northern Rivers Regional Development Board.

Starosielski, N. (2011), '"Movements that are drawn": a history of environmental animation from *The Lorax* to *FernGully* to *Avatar*', *The International Communication Gazette*, 73: 1–2, pp. 145–63.

TakingITGlobal (2011), 'deForestAction', http://gg.tigweb.org/tig/deforestaction. Accessed October 5, 2011.

Thomashow, M. (2002), *Bringing the Biosphere Home: Learning to Perceive Environmental Change*, Cambridge, Mass: RMIT Press.

Turner, G. (2006), 'The mass production of celebrity: "Celetoids", reality TV and the "demotic turn"', *International Journal of Cultural Studies*, 2006: 9, p. 153.

Virgo Productions (2011), '100 Days in the Jungle: what can twelve young people, a daring scientist, and millions of on-line supporters achieve in 100 days to save Planet Earth?', *http://www.anactionmovie.com/wp-content/uploads/exec_summary_july_2011.pdf*. Accessed October 5, 2011.

Vivanco, L. A. (2004), 'The work of environmentalism in an age of televisual adventures', *Cultural Dynamics*, 16: 5, pp. 5–27.

Webb, G. (2011), 'Interview conducted by Susan Ward', at Eureka, NSW, 18 May.

Willoquet-Maricondi, P. (2010), 'Shifting paradigms: from environmentalist films to ecocinema', in P. Willoquet-Maricondi (ed.), *Framing the World: Explorations in Ecocriticism and Film*, Charlottesville VA and London: University of Virginia Press.

Wilson, H. (2003), 'Introduction', in H. Wilson (ed.), *Belonging in the Rainbow Region: Cultural Perspectives on the NSW North Coast'*, Lismore: Southern Cross University, pp. 1–11.

Woods, M. (2007), 'Engaging the global countryside: globalisation, hybridity and the reconstitution of rural place', *Progress in Human Geography*, 31: 4, pp. 485–507.

YouTube (2013), 'Search for 10 people to star in a 3D movie and help save the planet!', http://www.youtube.com/watch?v=x_iRpMpPZs4. Accessed January 21, 2013.

Afterword–Towards a Transnational Understanding of the Anthropocene

Tommy Gustafsson and Pietari Kääpä

> That so many of us are here today is a recognition that the threat from climate change is serious, it is urgent, and it is growing. Our generation's response to this challenge will be judged by history, for if we fail to meet it – boldly, swiftly, and together – we risk consigning future generations to an irreversible catastrophe.

President Barack Obama opened his speech on climate change at the United Nations in 2009 in this way, authorizing the fact that climate change was an urgent issue that involved and concerned the entire globe. Since then, the 2010 United Nations Climate Change Conference was held in Cancún, Mexico, where a Green Climate Fund was proposed, a fund worth $100 billion a year by 2020, to assist poorer countries in financing emissions reductions and adaptation. There was, however, no agreement on how to extend the Kyoto Protocol, or how the finances for the Green Climate Fund would be raised. Nor did this agreement, or the ones made in Durban 2011 and Bangkok 2012, declare whether developing countries should have binding emissions reductions or whether rich countries would have to reduce their emissions first (UNFCCC 2012).

In fact, the successful mediation of awareness of climate change on a popular global scale, which in a way began with Al Gore's *An Inconvenient Truth* (2006) and continued with Obama's speech, seems to have come to a backlash. Both as a consequence of the international financial crisis of 2008, considered by many economists to be the worst financial crisis since the Great Depression, but also due to the search of the global media industries, in general, for more high profile material that can generate bolder headlines than yet another international inadequate agreement on the environment. Climate change appears to have left the main stage. The lack of interest and money has, as a consequence, left room in the spotlight for the adversaries of climate change who mainly argue two perspectives. The first one concerns the claim that 'we', or the contemporary national state or corporate organization, cannot afford climate improvement investments. The second declares that these investments are not necessary anyway since 'nobody' knows who is responsible for climate change—man or nature (Bell 2012; Fagerström 2011; Pugliese and Ray 2011; Angus Read Public Opinion 2011).

A common assertion binding the climate change deniers is that climate change reporting is exaggerated, with the media amplifying statistics in order to create, not awareness, but sensation (Stiernstedt 2009). Support for these assertions comes from many places. For example, Göran Ahlgren, a Swedish commentator on climate change and professor of organic chemistry, has argued that unfair climate reporting in the media can lead to ignorance. That is, exaggeration of statistics can contribute to wrong decisions being made by politicians, but it can also lead to information overload and fatigue among the public, and ultimately to the perception that the pressing ecological questions no longer appear to be as urgent (Ahlgren 2012).

Academic collections on ecocritical media have taken it as their mandate to respond to these sorts of metacommentaries on media reportage of climate change. Subtitles for collections on ecocinema also enforce the sense that films are cultural products responding to urgent and perilous issues affecting all life on earth. Lu and Mi's collection, for example, talks about 'environmental challenge' (2009), whereas this book proposes to discuss 'ecological transformation'. Both cases make it explicitly clear that the environment is in peril resulting explicitly from man-made changes. Most academic work in the ecocritical humanities takes this as a logical basis for their explorations. And so it is with this collection. Whether we are discussing ecocide or climate change, destruction of biodiversity or the inhumane treatment of plants or animals, the contributors to this volume are concerned with the ways human activity positions itself in a preordained superiority in relation to the rest of the environment.

While we could question the rationality of any academic approach that unquestionably accepts a politicized position such as this, activist aspirationality remains a key aspect of contemporary ecocinema studies much in a similar way that film criticism is obligated to uphold a connection with actual or virtual audiences in order to obtain credibility and social value (Brereton 2005: 235). Nonetheless, aspirationality and the perceptions of audiences can sometimes clash with the fact that narrative cinema often enlarges and simplifies portrayed events and facts in order to create, not awareness as the ecocritic would have it, but enough excitement to facilitate the maximum impact at the box office. Cinema is in this way analogous to the media's striving for newsworthiness and spectacle in general. Furthermore, narrative films do, almost by default, include an active subject or protagonist that, in the case of ecocinema, often equals man-made disturbance and intrusion on the environment. Thus, nature becomes inevitably objectivized in most films, ranging from Hollywood blockbusters (*The Day After Tomorrow*, 2004), to Chinese mainstream cinema (*Tianxia wuzei/A World without Thieves*, 2004) to documentaries (*Wuyong/Useless*, 2007).

One reason for this somewhat uncritical attitude to the beneficiality of adopting an ecocritical stance lies in the increasing acceptance and adoption of terms like the Anthropocene, a term for the contemporary era of our planet's existence. The term, according to Paul J. Crutzel and Christopher Schwägerl, indicates 'the age of men', an era different from the previous 12000 years where human influence was slowly integrated into the ecosystem, called the age of the Holocene. Whereas the many reasons for species extinction in the Holocene are consistently contested by scientists, the Anthropocene makes explicit the role humanity plays in this equation (Crutzel and Schwägerl 2011). The influence of humans is considered to be of a distinctly negative variety as much of the discussion revolves around the fundamentally destabilizing effect that humanity has on the sustainability of the planetary ecosystem. From the articles in this collection, we see a range of examples in which this destabilization takes place in the ecosystem. Consider our cover example, *Sanxia haoren*/*Still Life* (Jia Zhangke, 2006). Progress, modernization, development and sustainability intertwine to provide an exploration of not only how human activities impact natural systems, but also how they generate a general sense of malfare and disorientation. The instability generated by processes of capitalist sustainable development has a destructive impact that extends far beyond the local to what must be seen as a planetary issue. The rights and wrongs of the situation are up to debate, but what ultimately lingers is the sense that an Anthropocenic world is a world gone wrong, or life out of balance as Godfrey Reggio's seminal *Koyaanisqatsi* so effectively summarized already in 1983.

Most articles in this collection evoke a common problem in ecocritical studies involving the position of the analyst as a member of the human race. They are thus a potential contributor in the anthropogenic destabilization of the environment and part of the destructive logic of the Anthropocene—or a participant in diverting these practices to a more sustainable pathway. Lawrence Buell for one makes the suggestion that being part of what you criticize is an inevitable fallacy of ecocritisism and one that we just have to live with (Buell 2005). Rather than focusing on our embeddedness in the structures we criticize, we can turn attention to the ways diversifying our approach beyond grappling with questions of how anthropological methods of analysis can benefit ecocritical studies. One of the surest ways of attempting to engage with the Anthropocene in any satisfactory way is to diversify our methods of analysis. For this, transnational approaches prove especially useful. However, the transnational does not automatically equal a democratic approach to engaging with the Anthropocene through ecocriticism. Zygmunt Bauman's notion of democracy, for example, as 'a living, breathing network of many dialogues grounded in differences, differences which signify, and from whose interaction new conditions emerge, which have to be renegotiated in an unending debate', is idealistic (Bauman referenced in Cubitt 2005: 135). But is this form of democracy achievable on a global scale?

Often advocates of ecocriticism or critical approaches to ecocinema try to find a common denominator in the fact that 'we' all live on and share the same planet. For example, Pat Brereton suggests that we must 'appreciate the planet as a total unifying eco-system' (2005: 234) or, more commonly, by tracing back this unity to the famous photograph of the 'blue' Earth taken during the American lunar mission in 1972 (for example, Cubitt 2005: 133). But are these iconic conceptions really enough for the creation of even a notion of global unity and planetary democracy? The important but fragile agreements on a Green Climate Fund and emissions reductions of greenhouse gases are, we would argue, on the contrary sharp examples of the division of the world into richer and poorer nation-states, where green politics is 'a luxury for those who can afford to think ahead to the future' (Cubitt 2005: 138).

Cubitt declares, and rightly so, that the middle classes of the developed world 'are not likely to give up either their riches or their addiction to carbon fuels'. But he also suggests that education is a necessary ingredient for the advancement of ecological knowledge on a global scale, and that in turn 'demands the creation of a political class who can carry their arguments to the world stage. The predicament, however, is that 'by the nature of things', this class has the tendency to 'become increasingly remote from the population they represent' (Cubitt 2005: 138). The solution to this ecological conundrum is that there is no easy solution other than to create awareness and an understanding of the world that includes as many spheres of activities as possible—economic, political, environmental, geographical, education, entertainment, and so on. In accordance with this the transnational component in ecocinema studies is thus a further attempt to create an understanding that does not shy from cultural differences and the often wide gap between the local and the global.

Cultural diversity is an important aspect of this collection and one where it aims to make its most significant contribution to ecocinema studies. If the Anthropocene is the reality in which we must operate, we can try to unravel its implications for cultural approaches to ecocriticism by exploring the multiple ways in which anthropogenic climate change seems different from specific cultural perspectives. We can also complicate any notion of a simplistic or homogeneous idea of the Anthropocene by exploring the imbalances of power and exploitation that characterize it. Transnational approaches to the Anthropocene act as an important reminder at this late stage of our collection to both diversify our vocabulary and expand our range of analysis. Even when discussing the destruction caused by pan-human concerns on planetary ecosystems, we must be prepared to consider not only the ways in which we are part of what we criticize, but also how we can make some contribution to a situation that seems entirely enmeshed in skepticism and disillusionment. It is to this discussion that we hope our collection contributes.

References

Ahlgren, G. (2012), 'Mediernas osakliga klimatrapportering leder till okunskap', *Newsmill.se*, http://www.newsmill.se/artikel/2012/07/03/mediernas-osakliga-klimatrapportering-leder-till-okunskap. Accessed, September 2012.

Angus Read Public Opinion (2011), 'Britons question global warming more than Americans and Canadians. Half of respondents in the two North American countries think climate change is a fact and is caused by emissions—fewer Britons concur', http://www.angus-reid.com/wp-content/uploads/2011/09/2011.09.12_Climate.pdf. Accessed September 12, 2012.

Bell, L. (2012) 'Global warming alarmism: when science IS fiction', *Forbes.com*, http://www.forbes.com/sites/larrybell/2012/05/29/global-warming-alarmism-when-science-is-fiction/. Accessed September 12, 2012.

Brereton, P. (2005), *Hollywood Utopia. Ecology in Contemporary American Cinema*, Portland and London: Intellect.

Buell, L. (2005), *The Future of Environmental Criticism: Environmental Crisis and Literary Imagination*, MA Malden: Blackwell.

Crutzel, P. J. and Schwägerl, C. (2011), 'Living in the Anthropocene: toward a new global ethos', *Environment 360*, http://e360.yale.edu/feature/living_in_the_anthropocene_toward_a_new_global_ethos/2363/. Accessed September 14, 2012.

Cubitt, S. (2005), *Eco Media*, Amsterdam and New York: Rodopi.

Fagerström, J. (2011) 'Klimatpolitiken kommer att kosta oss vår välfärd', *Newsmill.se*, http://www.newsmill.se/artikel/2011/12/07/klimatpolitiken-kommer-att-kosta-oss-v-r-v-lf-rd. Accessed September 12, 2012.

Lu, S. and Mi, J. (2009) *Chinese Ecocinema: In an Age of Environmental Challenge*, Hong Kong: Hong Kong University Press.

Pugliese, A. and Ray, J. (2011), 'Fewer Americans, Europeans view global warming as a threat', http://www.gallup.com/poll/147203/Fewer-Americans-Europeans-View-Global-Warming-Threat.aspx. Accessed September 12, 2012.

Stiernstedt, J. (2009), 'En av tre svenskar tycker avtal är viktigt', *Dagens Nyheter*, December 7, 2009.

UNFCCC (2012), 'Cancun agreements', http://unfccc.int/meetings/cancun_nov_2010/items/6005.php. Accessed September 9, 2012.

Contributor Details

Kiu-wai Chu is a Ph.D. candidate in Comparative Literature from the University of Hong Kong. He received his Bachelor of Arts in Politics and South East Asian Studies from SOAS, University of London, and Master of Philosophy in Screen Media & Cultures from the University of Cambridge. He has been a lecturer in the Centre for International Degree Programmes (CIDP) of HKU SPACE from 2007 to 2010, teaching programmes in Asian Studies, Media and Cultural Studies. He has also recently been awarded a Fulbright scholarship to visit the University of Nevada, Reno (UNR) for ten months starting from September 2012. His research topic lies in Chinese cinema and visual culture, Ecocriticism, and Chinese environmental thoughts.

Ilda Teresa de Castro is completing her Ph.D. dissertation in Film Studies in the Department of Communication Sciences at the Faculty of Social Sciences and Humanities of the New University of Lisbon, in Portugal. Her thesis deals with the part films play in the construction of an ecocritical conscience.

Rebecca Coyle is an associate professor in the School of Arts and Social Sciences at Southern Cross University, Australia, and editor of *ScreenSound: The Australasian Journal of Soundtrack Studies* (see http://www.screensoundjournal.org/). Her publications deal with media representation, screen music and culturalstudies with a particular focus on Australian cultural production (see http://works.bepress.com/rebecca_coyle/). Her most recent research projects cluster around ecomedia, regional creativeindustries and screen animation. Email: rebecca.coyle@scu.edu.au.

Inês Crespo is a Ph.D. researcher. She has worked at the Joint Research Centre (JRC, European Commission, Italy) for the last 4 years with a fellowship with the New University of Lisbon. Her research aims to assess the role of films and documentaries in communicating climate change and constructing public perceptions about that project. She has a background in Environmental Engineering, having obtained a 5-year degree in 2006 at the University of Aveiro, Portugal.

Roberto J. Forns-Broggi has taught literature in Peruvian high schools and universities until 1991. In 1995, he obtained a Ph.D. in Spanish at Arizona State University with a dissertation about Vertical Poetry by Roberto Juarroz. In 2006, his short film, *The House of Wisdom*, was shown at the 29th Starz Denver Film Festival. He is doing research on Latin American cinema and literature. Currently he is a full Professor of Spanish at Metropolitan State University of Denver. A collection of his essays on ecological perspectives applied to literature, art, film and new media, 'Abecedario de la imaginación ecológica en nuestras Américas' [*Handbook of Ecological Imagination in Our Americas*] will be published in Lima this year.

Tommy Gustafsson holds a Ph.D. in History and is an Associate Professor of Film Studies at Linnæus University in Sweden. His research focuses on Swedish silent film with a primary focus on the representations of gender, sexuality and racial stereotyping, for example in an article on Oscar Micheaux and Swedish film culture published in *Cinema Journal*. He has also written about the relationship between film/television and history, including how the Rwandan genocide was portrayed and explained via feature films and television news. Currently he is working on a project that investigates how Children's Television was used to teach history to children in Sweden in the 1970s. He is the co-editor of a special issue of the journal *Interactions*, 'Ecocinemas of transnational China'.

Pietari Kääpä is a research fellow in the School of Film and Television Studies at University of Helsinki. His research focuses on transnational cinema in a range of cultural contexts and theoretical frameworks. In addition to multiple articles, he has published two monographs and two collections on the globalization of Finnish cinema. He is currently working on several projects focusing on ecocinema, including editing special issues of the journal *Interactions* ('Ecocinemas of transnational China'; 'The audiences of ecocinema'), and working on a monograph on Nordic ecocinemas.

TAM Yee Lok is a senior research assistant of the Hong Kong Advanced Institute for Cross-Disciplinary Studies of City University of Hong Kong. He has graduated from the Hong Kong University of Science and Technology with the degrees of Master of Art and Master of Philosophy in Humanities in 2007 and 2009, respectively. His research interests are ecocriticism, Sinophone literature and global Chinese cinemas and his dissertation title is 'Resurgence of (inter)connectivity: An ecocritical approach to nature, animal and body in Hong Kong literature'.

Corrado Neri holds a doctorate in Chinese Film Studies from the University of Ca' Foscari, Venice, and the University Jean Moulin, Lyon 3. He is now assistant professor at the Jean Moulin University, Lyon 3. He has conducted extensive research on Chinese cinema in Beijing and Taipei and published many articles on books and magazine (in English, French and Italian). His book *Tsai Ming-liang* on the Taiwanese film director appeared in 2004 (Venezia, Cafoscarina). His second book, *Ages Inquiets. Cinémas chinois: une representation de la jeunesse*, was published in 2009 (Lyon, Tigre de Papier). He co-edited (with Kirstie Gormley) a bilingual (French/English) book on Taiwan cinema (*Taiwan cinema/Le Cinéma taiwanais*, Asiexpo, 2009), and *Global Fences* (with Florent Villard, IETT, 2011).

Ângela Guimarães Pereira has a Ph.D. in Environmental Assessment with the New University of Lisbon. Since 1996 she is working at the Joint Research Centre of the European Commission, where with others she formed the Knowledge Assessment Methodologies group. She is responsible for activities on science and society interfaces that range from knowledge quality assurance methodologies to social research and deployment of new information technologies for EU inclusive governance prospects. She has also organized in 2003 the first landmark conference on Interfaces between Science and Society in Milan.

Susan Ward is a researcher with Southern Cross University. Her interest is in Australian film and television and the spatial dynamics of audio-visual production. She is co-author of *Local Hollywood: Global Film Production and the Gold Coast* (with Ben Goldsmith and Tom O'Regan). More recently her focus has taken on a more environmental hue looking at the impact of climate change on the cultural industries especially in relationships between place, cultural production and representative practices in environmentalism.

Index